ECONOMIC ISSUES, PROBLEMS AND PERSPECTIVES

STRATEGIES FOR AMERICAN INNOVATION

Economic Issues, Problems and Perspectives

Additional books in this series can be found on Nova's website under the Series tab.

Additional E-books in this series can be found on Nova's website under the E-books tab.

ECONOMIC ISSUES, PROBLEMS AND PERSPECTIVES

STRATEGIES FOR AMERICAN INNOVATION

ZACHARY A. MONROY
EDITOR

Nova Science Publishers, Inc.
New York

For permission to use material from this book please contact us:
Telephone 631-231-7269; Fax 631-231-8175
Web Site: http://www.novapublishers.com

Library of Congress Cataloging-in-Publication Data

Strategies for American innovation / editor, Zachary A. Monroy.
p. cm.
Includes index.
ISBN 978-1-61324-247-6 (hardcover)
1. Technological innovations--Economic aspects--United States. 2. Economic development--United States. I. Monroy, Zachary A. II. Title.

HC110.T4S78 2011
338'.0640973--dc22

2011010162

Published by Nova Science Publishers, Inc. New York

CONTENTS

PREFACE

America's future economic growth and international competitiveness depend on our capacity to innovate. This book examines how the Administration, the American people and American businesses can work together to strengthen our long-run economic growth. It explains the essential role of innovation in our past and future prosperity, the importance of the private sector as the engine of innovation, and the role of government in supporting our innovation system.

Chapter 1- America's future economic growth and international competitiveness depend on our capacity to innovate. We can create the jobs and industries of the future by doing what America does best – investing in the creativity and imagination of our people. To win the future, we must out-innovate, out-educate, and out-build the rest of the world. We also must take responsibility for our deficit, by investing in what makes America stronger, cutting what doesn't, and reforming how our government operates so it focuses on promoting our economic growth and preparing for the challenges of a new century.

Chapter 2- Since taking office, President Obama has taken historic steps to lay the foundation for the innovation economy of the future. The Obama Innovation Strategy builds on well over $100 billion of Recovery Act funds that support innovation, additional support for education, infrastructure and others in the Recovery Act and the President's Budget, and novel regulatory and executive order initiatives. It seeks to harness the inherent ingenuity of the American people and a dynamic private sector to ensure that the next expansion is more solid, broad-based, and beneficial than previous ones. It focuses on critical areas where sensible, balanced government policies can lay the foundation for innovation that leads to quality jobs and shared prosperity. It has three parts:

Chapter 3- President Obama understands the importance of innovation for sustainable growth and quality jobs. On September 21st, 2009, he released his *Strategy for American Innovation* that identified three critical roles for the Federal Government: to invest in the building blocks of innovation; to create the right environment for private sector investment and competitive markets by, for example, promoting high-growth entrepreneurship, protecting U.S. intellectual property rights, and fostering an open government; and to serve as a catalyst for breakthroughs related to national priorities such as clean energy, health care, and other "grand challenges" of the 21st century.

In my capacity as Assistant to the President, Chief Technology Officer, and Associate Director for Technology in the Office of Science and Technology Policy, my mission is to harness the power and potential of technology, data, and innovation to transform the Nation's economy and to improve the lives of everyday Americans. The Administration envisions an economy in which jobs are more plentiful, American firms are more competitive, Americans are safer and more secure, and energy use is cleaner and more economical.

Strategies for American Innovation
Editor: Zachary A. Monroy

ISBN: 978-1-61324-247-6

Chapter 1

A STRATEGY FOR AMERICAN INNOVATION: SECURING OUR ECONOMIC GROWTH AND PROSPERITY*

National Economic Council, Council of Economic Advisers, and Office of Science and Technology Policy

EXECUTIVE SUMMARY

America's future economic growth and international competitiveness depend on our capacity to innovate. We can create the jobs and industries of the future by doing what America does best – investing in the creativity and imagination of our people. To win the future, we must out-innovate, out-educate, and out-build the rest of the world. We also must take responsibility for our deficit, by investing in what makes America stronger, cutting what doesn't, and reforming how our government operates so it focuses on promoting our economic growth and preparing for the challenges of a new century.

President Obama's *Strategy for American Innovation* seeks to harness the inherent ingenuity of the American people to ensure that our economic growth is rapid, broad-based, and sustained. Innovation-based economic growth will bring greater income, higher quality jobs, and

---

* This is an edited, reformatted and augmented version of a National Economic Council's publication, dated FEBRUARY 2011.

improved health and quality of life to all U S citizens. The *Strategy for American Innovation* provides a multifaceted, commonsense, and sustained approach to ensuring America's future prosperity.

This document updates the *Innovation Strategy* issued in September 2009, detailing how the Administration, the American people, and American businesses can work together to strengthen our long-run economic growth. It begins by explaining the essential role of innovation in our past and future prosperity, the central importance of the private sector as the engine of innovation, and the role of government in supporting our innovation system. Building from this framework and the initiatives set forth in the first innovation strategy document, it charts the progress of our initial efforts, discusses additional steps implemented in the past year, and introduces important new initiatives, all captured by the pyramid below.

New Initiatives

Key Administration priorities will improve America's economic growth and competitiveness on many critical dimensions.

- The Administration's proposed **Wireless Initiative** will help businesses reach 98% of Americans with high-speed wireless access within five years and also facilitate the creation of a nationwide interoperable public safety network. The Initiative will substantially expand the development of new wireless spectrum available for wireless broadband, by freeing up 500 MHz over 10 years. Expanding new commercial spectrum is necessary to avoid "spectrum crunch" and facilitate the rapidly growing wireless technology revolution. The initiative will support advances in security, reliability, and other critical wireless features; accelerate wireless innovations in health, education, transportation, and other application areas; and engage community participation in generating and demonstrating next generation wireless applications.
- The **patent reform** agenda is essential to reducing the enormous backlog of patent applications at the U.S. Patent & Trademark Office (USPTO). By stalling the delivery of innovative goods and services to market, this backlog impedes economic growth and the creation of high-paying jobs. The patent reform legislative agenda will enable the USPTO to adequately fund its operations through user fees and allow the agency to implement new initiatives to improve patent quality while reducing the average delay in patent processing times from 35 months to 20 months. Once implemented, the USPTO's proposed three-track model will allow applicants to prioritize applications, enabling the most valuable patents to come to market within 12 months.
- The Administration is developing new initiatives to improve **K-12 education** with an emphasis on graduating every student from high school ready for college and a career. The Administration's FY 2012 Budget will launch the Advanced Research Projects Agency – Education (ARPA-ED) to support research on breakthrough technologies to enhance learning. The Budget also supports continuation of the historic Race to the Top, with an expanded focus on school districts prepared to implement and sustain comprehensive reforms. Working with a coalition of private sector leaders called Change the Equation, the Administration is encouraging public-private partnerships that inspire more students – including girls and other currently underrepresented groups – to excel in science, technology, engineering, and mathematics (STEM). The

Administration will also work to prepare 100,000 STEM teachers over the next decade with a down payment in the FY 2012 Budget to recruit STEM teachers and improve teacher training.

- To accelerate the development of **clean energy** technologies, the President has proposed a Clean Energy Standard that will help us reach a goal of delivering 80% of the nation's electricity from clean sources by 2035. The Administration's FY 2012 Budget proposes to expand the funding to date for the Advanced Research Projects Agency – Energy (ARPA-E) and to create three new Energy Innovations Hubs to solve challenges in critical areas. The Budget also proposes a reauthorization of the Clean Energy Manufacturing Tax Credit and provides funding for research, development, and deployment to help the U.S. reach the goal of one million advanced technology vehicles on the road by 2015.
- The ***Startup America*** initiative will facilitate entrepreneurship across the country, increasing the success of high-growth startups that create broad economic growth and quality jobs. The Administration launched the *Startup America* initiative with new agency efforts that accelerate the transfer of research breakthroughs from university labs; create two $1 billion initiatives for impact investing and early-stage seed financing, among other incentives to invest in high-growth startups; improve the regulatory environment for starting and growing new businesses; and increase connections between entrepreneurs and high-quality business mentors. Responding to the President's call to action around the national importance of entrepreneurship, private-sector leaders are independently committing significant new resources to catalyze and develop entrepreneurial ecosystems across the country.

These new initiatives work together with, and expand upon, sustained Administration efforts across our three critical areas. These ongoing efforts are summarized as follows:

1. Invest in the Building Blocks of American Innovation. Spurring the innovations that will drive America's future economic growth and competitiveness requires critical investments in basic foundations: our workforce, our scientific research, and our infrastructure.

- **Educate the next generation with 21st century skills and create a world-class workforce.** President Obama is taking continuous steps to improve our educational system from early education through college and graduate school, and to promote student achievement and careers in STEM fields. In early childhood, the Administration is supporting innovation through the Early Learning Challenge Fund and injecting performance-based competition into the Head Start Program. At the elementary and secondary levels, the Educate to Innovate campaign harnesses public-private partnerships to enhance STEM education, complementing continuing efforts such as Race to the Top, which uses competitive grants to leverage state and local reform. At the college level and beyond, the Administration is committed to restoring America's global leadership in college graduation rates (by, among other strategies, improving affordability through the Student Aid and Fiscal Responsibility Act), making investments in community colleges and the public workforce system, and supporting the new Task Force on Skills for America's Future, which will leverage public-private partnerships to better train Americans of all ages for the jobs of today and tomorrow.

- **Strengthen and broaden American leadership in fundamental research.** The commercial innovations that drive economic progress often depend on breakthroughs in fundamental science. President Obama has implemented the largest increase in federally-funded research in history and is making continuous investments to double funding for three key basic research agencies: the National Science Foundation, the Department of Energy's Office of Science, and the National Institute of Standards and Technology laboratories. These sustained science investments will lay the foundation for new discoveries and new technologies that will improve our lives and create the jobs and industries of the future. These investments will help the United States establish a leadership position in areas such as robotics and data-intensive science and engineering.
- **Build a leading 21st century infrastructure.** President Obama has made a renewed commitment to investing in the roads, rails, and runways that America's businesses need to be efficient and innovative. Building on historic investments through the Recovery Act, the Administration continues to address transportation challenges through investment in high-speed rail, the next generation of air traffic control, and a new proposal for a National Infrastructure Bank, which will promote competition and innovation to maximize the return on our infrastructure investments.

- **Develop an advanced information technology ecosystem.** President Obama has developed a comprehensive strategy to create the IT ecosystem needed for 21st century innovation. This "virtual infrastructure" encompasses the critical information, computing, and networking platforms that increasingly support our national economy. With constant effort and vigilance, the Administration is working to expand access to high-speed Internet, modernize the electric grid, increase the availability of wireless spectrum to support high value uses, and secure cyberspace.

2. Promote Market-Based Innovation. American businesses are our engines of innovation. They bring American ingenuity to the marketplace, where new ideas are proven, commercialized, and diffused. It is imperative to promote a national environment ripe for innovation and entrepreneurship that allows U.S. companies to drive future economic growth and continue to lead on the global stage.

- **Accelerate business innovation with a simplified and permanent R&E tax credit.** President Obama has called for the Research and Experimentation Tax Credit to be simplified and made permanent, creating predictable, substantial incentives for U.S. businesses to innovate. The proposed FY 2011 Budget devotes about $100 billion over 10 years to leverage additional research and development investments.
- **Support innovative entrepreneurs.** President Obama has expanded lending support and tax credits for small businesses while supporting well-functioning capital markets for businesses of all sizes. In addition to patent reform, which will accelerate patent issuance and better enable new companies to succeed, and the *Startup America* initiative, which will promote entrepreneurship across the country, the Affordable Care Act removes obstacles to entrepreneurship by making it easier for Americans to start and join new businesses without giving up health coverage.
- **Catalyze innovation hubs and encourage development of entrepreneurial ecosystems.** President Obama continues to emphasize the potential of "innovation hubs," looking for new opportunities to bring talented scientists and entrepreneurs together to support innovation in cutting edge areas. This concept underlies the Department of Energy's Energy Innovation Hubs program and is also

driving the *Startup America* initiative's focus on building connections between established and new entrepreneurs, including those making the leap from lab to industry. Through these efforts, the Obama Administration is working to catalyze a new model of economic development.

- **Promote innovative, open, and competitive markets.** President Obama is working to encourage innovation by improving regulation and enhancing market access at home and abroad. The revised Horizontal Merger Guidelines, released in August 2010, bring innovation considerations forcefully into antitrust evaluation. In addition, through efforts such as the free trade agreement with South Korea, the National Export Initiative brings a sustained, vigorous commitment to ensure fair and open export markets for American producers, allowing our innovative businesses to expand globally with the goal of doubling exports by the end of 2014.

3. Catalyze Breakthroughs for National Priorities. For national priorities where innovation is critical but market failures impede progress, government can help spur technological advances. Priorities include developing alternative energy sources, reducing costs and improving care with health IT, catalyzing advances in educational technologies, and ensuring that the U.S. remains on the leading edge of the bio- and nanotechnology revolutions.

- **Unleash a clean energy revolution.** New and improved energy technologies will play central roles in the 21st century global economy, and the Obama Administration is committed to fostering American leadership in this area, providing economic growth and creating jobs of the future while confronting environmental challenges and enhancing our energy security. Building on the successful investments geared toward doubling the supply of renewable energy by the end of 2012, and other initiatives, the Administration is setting ambitious new goals and laying the pathways to meet them. Through the proposed Clean Energy Standard, expanded investments in research through the Department of Energy's Office of Science, ARPA-E, three new Energy Innovation Hubs, and other means to accelerate research, development, and deployment of clean energy technologies, the Administration will shift the American economy toward global leadership and a clean, secure, and independent energy future.

- **Accelerate biotechnology, nanotechnology, and advanced manufacturing.** The President is committed to investments in innovation that promise to drive better health, future economic growth, and quality jobs in America. The National Institutes of Health has proposed a new National Center for Advancing Translational Sciences, which will speed the development of new diagnostics, treatments, and cures by building new bridges between the lab and clinic. The National Nanotechnology Initiative (NNI) is investing in areas such as nanoelectronics, which will foster a revolution in computing comparable to the transition from the vacuum tube to the transistor. And the FY 2012 Budget is making substantial investments to accelerate breakthroughs in advanced manufacturing technologies that can provide foundations for private sector investment and growth.
- **Develop breakthrough space capabilities and applications.** Space capabilities play critical roles in global communications, navigation, and commerce, while warning of natural disasters and improving national security. Guided by the National Space Policy, NASA, the Department of Defense, and other agencies are working to advance U. S. capabilities and expand American industry's role in developing next-generation applications.
- **Drive breakthroughs in health care technology.** Innovations in health care delivery, harnessing the power of data and technology, promise to help prevent medical errors, improve care quality, and reduce costs. Building from the Recovery Act and the Affordable Care Act, the Administration is continuously engaged in projects to promote health IT adoption, reform payment incentives to reward value instead of volume, and liberate an unprecedented amount of health information. In combination, these trends will facilitate fundamental improvements in national health and harness American ingenuity in solving our health care challenges.
- **Create a quantum leap in educational technologies.** The United States should foster innovation in technologies that have the potential to dramatically improve student performance, such as software that is as effective as a personal tutor, and increase access to lifelong learning and training for American workers. The President's FY 2012 Budget for the Department of Education includes a proposal to launch the Advanced Research Projects Agency – Education, a new

organization that will support research on breakthrough technologies to enhance learning.

INTRODUCTION

"For in a global economy, the key to our prosperity will never be to compete by paying our workers less or building cheaper, lower-quality products. That's not our advantage. The key to our success – as it has always been – will be to compete by developing new products, by generating new industries, by maintaining our role as the world's engine of scientific discovery and technological innovation. It's absolutely essential to our future."

— President Barack Obama, November 17, 2010

Innovation Is Essential to Winning the Future through Long-Term Growth and Competitiveness

The history of the American economy is one of enormous progress associated with remarkable innovation

Two hundred years ago, real income per person in America averaged four percent of what it is today,' the average American lived for forty years, and thirty percent of children did not survive until their fifth birthday.[2] Electric power, automobiles, and telephones were hardly imagined, let alone computers and air travel. There were no antibiotics or vaccines – and no understanding that germs cause disease. The word "scientist" had not yet been coined. But researchers like Isaac Newton had begun uncovering fundamental scientific foundations that would underpin two centuries of practical inventions. The U.S. Constitution empowered Congress to create effective intellectual property rights – helping add "the fuel of interest to the fire of genius," in President Lincoln's words. Americans later seized on the Industrial Revolution – an explosion of innovation – propelling a young country with democratic ideals to unprecedented economic heights and providing a powerful example for other nations to follow. In short, innovation is ultimately tied to America's well-being and to our conception of the essential "American character."

Innovation – the process by which individuals and organizations generate new ideas and put them into practice – is the foundation of American economic growth and national competitiveness. Economic growth in advanced economies like the United States is driven by the creation of new and better ways of producing goods and services, a process that triggers new and productive investments. That innovation is the cornerstone of economic growth can be seen in the advance of our national industries.

Entire industries were made possible only by developing and commercializing new ideas, from the 19th century advances in railways and steam power, to the later revolution of electrification and the associated development of light bulbs, radios, televisions, electric refrigeration, and air conditioners, to the modern semiconductor, computer, and biotechnology industries. These innovative sectors have consistently raised the output of our workforce, creating better-paying jobs, raising our national standard of living, and enhancing our economic strength vis-à-vis other nations. Innovation can take many forms: a new machine that improves quality and production time in factories; a new consumer electronic device or Internet-enabled application that keeps us connected with coworkers and family; a new way of organizing the workplace that increases our productivity; or a new vaccine that protects our citizens from disease.

Since the 1940s, the United States has led the world in creating new industries and ways of doing business, establishing itself as the global innovation leader. But America cannot rest on its laurels. Unfortunately, there are disturbing signs that America's innovative performance slipped substantially during the past decade. Across a range of innovation metrics – including growth in corporate and government R&D, the number of scientific and technical degrees and workers, access to venture capital, and the creation of new firms – our nation has fallen in global innovation-ranked competitiveness.[3] Other nations recognize that innovation is the key to long-term economic growth and are making pro-innovation investments and adopting pro-innovation policies. Without thoughtful, decisive, and targeted actions, we cannot expect that the industries of the future will emerge and prosper in the United States.

Recognizing the central role of innovation in economic growth, the Administration's *Innovation Strategy* announced in 2009 emphasized several of these disturbing trends and called for renewed and enhanced investment in innovation. These efforts were substantially supported by historic investments in the Recovery Act and included large expansions in fundamental research through agencies such as the National Science Foundation, the Department of Energy's Office of Science, and the National Institutes of Health, accelerating fundamental breakthroughs at the beginning of the innovation pipeline. Now we must build on these efforts and ensure that the private sector can be as innovative as possible so that American workers and businesses will continue to lead the world economy in the decades ahead. New initiatives will free up wireless spectrum that will facilitate private sector investment and innovation, improve the patent system, train workers for quality jobs, catalyze the private sector to meet national priorities like clean energy, and foster the entrepreneurial spirit that has always driven this country to greater heights This *Strategy for American Innovation* discusses these new points of emphasis. and places them within the broader framework of the Administration's innovation policy.

Americans have always seen themselves as experimenters and risk-takers. Now we must – at every level of society – encourage this pioneering spirit. In the 1800s, when farmers and blacksmiths took hammers to plows and harnesses, America was described as a "nation of tinkerers." In the 21st century, continued economic growth depends on us being a "nation of innovators" – a nation that generates the best and brightest ideas and sees that these ideas spread through our workforce. The American people will do best when their inventive, entrepreneurial spirit is unleashed. Government policy must nurture that spirit and ensure it is not deterred.

The Private Sector Is America's Innovation Engine

America's entrepreneurs and industrial research laboratories have long produced a cascade of important innovations, from agricultural technologies to Edison's light bulb to Bell Labs' transistor, from General Electric's jet engines to Google's Internet tools. Innovation is not limited to new products but extends to new organizational models. Henry Ford's assembly line brought affordable automobiles to Americans while bringing higher employment and wages to the car industry. Dell Computer and Amazon.com similarly developed new sales models, harnessing the Internet to bring new, competitively-priced choices to a wider array of consumers.

America's businesses, with close knowledge and acute awareness of the costs and opportunities across our market system, are well positioned to tap the ingenuity of our workforce to solve specific challenges and cultivate new ideas in the crucible of competition. In so doing, they can perceive and generate commercially valuable ideas. And a new idea is just the starting point, because our market system, through its competitive pressures, also works to test these ideas and spread the best ones. Innovation is the entire process through which an invention is successfully put into practice and widely diffused, generating increased labor productivity for workers, profits for suppliers, and benefits to adopters and consumers.

Box 1: Social Gains from Innovation

The social gains from innovations typically greatly exceed the private return For example, the inventions of the telephone, transistor, light bulb, dishwasher, laser, CT scan, web browser, and antibiotics have all had enormous, broad, and ongoing social benefits far in excess of any commercial profits enjoyed by the original creators. General estimates suggest that the private profits from an innovation typically account for a tiny fraction – a few percent – of the social value.

Private firms do not capture the full gains of their innovations for three essential reasons. First, users will only pay for the innovation if its benefits exceed its price. These consumer benefits – the "consumer surplus" – mean that much of the innovation's value will immediately accrue to users. Second, the innovative business will face pressures to lower prices as other businesses imitate and improve upon the successful innovation, which shortens the profit stream of the original product; indeed, the firm that initially introduces a new innovation may not even be the one that succeeds in the marketplace. Moreover, once any intellectual property rights (the patent or copyright) expire, competitive pressures will further drive down prices and limit private profits, transferring the innovation's value even more fully to the consumer Finally, a successful innovation often launches hosts of additional innovations by other firms, the benefits of which are not captured by the original innovator.

With the limited scope for sustained profits in the private sector, the benefits of innovation accrue widely. Ultimately, innovation benefits society at large in the form of rising standards of living Real incomes rise, with Americans producing more output per hour and earning more income per hour, allowing us to consume new and improved products and live longer lives.

By demonstrating how ideas can be commercialized, businesses also drive other firms to innovate, allowing organizations with different technical capacities and market insights to take the next steps. In fact, most innovation does not stand alone but complements other innovations. For example, while Apple Computer founders Steve Jobs and Steve Wozniak envisioned a mass market for a more user-friendly personal computer, it took many years and many other contributions – including improvements in software, microprocessors, monitors, memory chips, batteries, and communications technology – before the personal computer industry reached critical mass. And these complementary technologies were themselves innovations, collectively providing jobs in engineering, manufacturing, distribution, and sales, while delivering significant consumer benefits.

The American economy is built on an enduring capacity for idea creation and diffusion. Competitive markets provide strong incentives for private businesses to improve their products and operations and for capital and labor resources to be reinvested in our best ideas. The inherent uncertainty of innovation means that important breakthroughs may come from many quarters – often unexpectedly – and our decentralized markets facilitate the generation of these new possibilities. By continually reinventing itself, the private sector is the engine of innovation that brings greater prosperity to Americans.

Government as Innovation Facilitator

Given the central importance of innovative activity to our economic growth, the public interest in sustaining innovation is clear The key follow-on question is whether markets alone can provide sufficient incentives for such investments The standard lesson from economics, and history, is that an innovation-friendly environment requires public support on specific dimensions [4] The appropriate role for government can be understood by clarifying the precise circumstances where markets, despite their many strengths, will not produce a sufficient stream of innovations on their own. Thus, the true choice in innovation policy is not starkly between government management and no government involvement, but rather choosing the right role for government in supporting private sector innovation.

One central "market failure" is in the field of basic scientific research. Basic research typically does not have direct commercial payoffs. Yet breakthroughs in basic research underpin downstream, commercial ideas, which can bring enormous economic benefits. For example, engineering builds on Newton's laws of motion, the biotechnology industry builds on Watson and Crick's discovery of the structure of DNA, and the dot.com industry builds on government and university development of the Internet. Because basic science has little if any immediate commercial return, its costs are typically not easily undertaken by private investors, thus leaving government funding as a critical source of support.

Other "market failures" surround commercial invention, where markets may still fail to provide adequate incentives. As discussed in *Box 1*, businesses typically capture only a small portion of the benefits of their innovations, partly because consumers enjoy a substantial share of the social gain and partly because follow-on innovations may be captured by other firms.[5] This general issue calls for policies that enhance private sector innovation incentives. These policies can act to lower commercial research costs through mechanisms such as the Research & Experimentation Tax Credit. Other mechanisms enhance the demand for innovations. For example, demonstration funding and government procurement can encourage the creation and deployment of next generation technologies, bringing private innovation incentives closer to the social interest. Government procurement was used, for example, by the Defense Department to promote the development of the Internet. Prize competitions (*see Box 2*) can be especially useful in driving innovation for specific needs. Collectively, these demand mechanisms can be targeted at well-defined national priority areas, such as clean energy, and can be especially useful in contexts where markets under-price an activity's costs, such as our national dependence on fossil fuel consumption.

Government also plays an essential role in setting and enforcing appropriate rules. Foremost in the innovation context is a well-functioning property rights system. Absent effective legal protections for innovators, other businesses can immediately exploit an innovator's idea, undermining the incentive to invent in the first place. Public policy solves this problem through intellectual property rights – allowing limited, short-run grants of exclusive rights to catalyze inventive activity. Recognizing the importance of intellectual property rights, we must commit to their effective enforcement, as the Obama Administration has done in appointing the first Intellectual Property Enforcement Coordinator and redoubling our efforts in this area. Critically, we must also commit to making the necessary public investments to support high-quality patent examination, lower legal uncertainty, and clear persistent patent application backlogs so that innovative businesses and entrepreneurs are not faced with unnecessary risk or left waiting for years before a patent decision is made.

Box 2: Spurring Innovation through Prizes and Challenges

In his September 2009 *Innovation Strategy*, President Obama called on all agencies to increase their use of prizes and challenges to mobilize America's ingenuity to solve some of our nation's most pressing challenges.

From the 1714 Longitude Prize that stimulated the development of the world's first practical method to measure a ship's longitude to the Orteig Prize that inspired Charles Lindbergh to fly nonstop from New York to Paris to the X Prize given to the inventor who creates a car that can deliver over 100 miles per gallon, prizes have a long record of spurring innovationIn the 21st century, unprecedented levels of connectivity have given rise to a renaissance for prize competitions. A recent McKinsey report found that private sector investment in prizes has increased significantly in recent years, including $250 million in new prize money between 2000 and 2007.

As the Wall Street Journal recently concluded, "These prizes have proliferated because they actually work." Under the right circumstances, prizes have a number of advantages over traditional grants and contracts. Prizes allow the sponsor to set an ambitious goal without selecting the team or approach that is most likely to succeed, to increase the number and diversity of minds tackling tough problems, to pay only for results, and to stimulate private-sector investment that is many times greater than the cash value of the prize (Orteig competitors collectively spent $400,000 to win the $25,000 purse).

Since September 2009, the Obama Administration has taken important steps to accelerate public-sector adoption of these innovative tools. In March 2010, the Office of Management and Budget issued a formal policy framework to guide agency leadership in using prizes to advance their core mission. In September 2010, the Administration launched Challengegov, a one-stop shop where entrepreneurs and citizen solvers can find public-sector prizes. Throughout, the Administration built a community of practice for agencies to share best practices and lessons learnedAs a result, in its first 4 months alone, Challengegov has featured nearly 60 challenges from more than 25 agencies across the Executive Branch, generating novel solutions for childhood obesity, advanced vehicle technologies, financing for small businesses, Type 1 Diabetes, and many other national priorities.

Another example is a challenge launched on the NASA Innovation Pavilion for a forecasting algorithm to protect America's astronauts from radiation exposure in space. Over 500 problem solvers from 53 countries answered NASA's callExpecting no solutions for this long intractable problem, NASA received a solu-tion that exceeded their requirements from a retired radio-frequency engineer in rural New Hampshire. The winner had never before responded to a government request for proposals, let alone worked with NASAYet, his winning approach forecast solar proton events with 85% accuracy, a result NASA dubbed "outstanding".

Prize competitions like these mark a dramatic departure from business as usual and – thanks to newly-enacted legislation – will soon become a standard tool in every federal agency's toolbox. On December 21st, Congress passed the America COMPETES Reauthorization Act of 2010, providing broad prize authority to all federal agenciesBy giving agencies a simple and

> clear legal path, the America COMPETES Reauthorization Act will make it dramatically easier for agencies to use prizes and challenges. In the months to come, the Obama Administration will work closely with key agencies to leverage the new authority for ambitious prizes in areas of national priority.

Government can also accelerate innovation by setting rules in specific sectors, notably by updating or eliminating outdated regulations. To that end, President Obama signed an Executive Order in January 2011 calling for a government-wide review of regulations to update or revise them to, among other purposes, ensure that they facilitate—rather than impede—competition and innovation in addressing valid public policy goals. Effective management of public resources, such as the electromagnetic spectrum, unleashes innovation by opening markets and reducing uncertainty over usage rights and engineering design. In appropriate contexts, public leadership can help set standards for technology platforms, such as emerging smart grid or health IT technologies, providing confidence to the marketplace to develop and adopt new generations of products. Standard setting, which the government can enable through its role as convener and support through research and development, often involves facilitating coordination within the private sector to create a larger market, thus enhancing the demand for innovative products. Export initiatives further increase the market scale for U S businesses. Increased scale is an attraction to business innovation, while tiny, balkanized markets are not.

Finally, the government plays essential roles through public investments that businesses rely on but do not themselves create. Educational investments and workforce training provide the essential building block – a capable workforce – from which new ideas come and through which good ideas spread. Infrastructure investments, including physical and information infrastructure, connect our markets, creating attractive scale for innovation and allowing best practices to diffuse.

By championing policies that facilitate marketplace innovation, the federal government will continue to be an essential partner in the U.S. national innovation system. To that end, the Obama Administration will take appropriate public action by supporting an environment in which innovation is rewarded and best practices are diffused, investing in a technically capable workforce, supporting basic scientific discoveries, and promoting the development of the technology platforms from which future innovations will spring. Government direction can never be a substitute for the free market conditions that propel American innovation. But government must act to support those conditions and ensure that innovation, the engine of our prosperity, drives America further and faster towards higher quality jobs, healthier and longer lives, new opportunities and new industries, and the ever-expanding technological frontier.

Invest in the Building Blocks of American Innovation

> "Competition [in the global economy] is going to be much more fierce and the winners of this competition will be the countries that have the most educated workers, a serious commitment to research and technology, and access to quality infrastructure like roads and airports and high-speed rail and high-speed Internet. Those are the seeds of economic growth in the 21st century. Where they are planted, the most jobs and businesses will take root."
>
> —President Barack Obama, December 6, 2010

Our innovation strategy begins with critical foundations: education, scientific research, and infrastructure. First, we must create an educational system that is internationally competitive and innovative in preparing our workforce for our increasingly knowledge-intensive economy. Next, we must invest in scientific research to restore America's leadership in creating the scientific and technological breakthroughs that underpin private sector innovations. Finally, we must invest in a first-class infrastructure that moves people and ideas at 21st century speeds. These are the building blocks of an innovation strategy that will lead America to a more prosperous future.

Educate Americans with 21st Century Skills and Create a World-Class Workforce

With strong educational foundations, Americans will create the leading ideas of the 21st century and ensure that these ideas diffuse throughout the American workforce. On many metrics, however, including grade-level proficiency and college graduation rates, America has slipped behind other countries. We must reform our education and workforce training systems to ensure Americans are qualified for the jobs of tomorrow. This imperative underpins the Obama Administration's focus on education reform in general and in science, technology, engineering, and math (STEM) education in particular. It is also imperative to extend STEM educational and career opportunities to women and minority groups that are underrepresented in these areas, so that all Americans can find quality jobs and lead our innovative economy in the decades ahead.

Improve America's Science, Technology, Engineering and Math (STEM) Education

The President has pledged to prepare an additional 100,000 STEM teachers by the end of the decade, with strong teaching skills and deep content knowledge. The Administration's Educate to Innovate campaign harnesses public-private partnerships to improve K-12 STEM education, make STEM education more accessible, move American students up the international rankings in STEM literacy, and expand STEM career opportunities. Responding to the President's call to action, over 100 CEOs have formed an organization called Change the Equation. In 2011, these business leaders have committed to scale up effective STEM programs in at least 100 under-served communities, with particular emphasis on broadening participation to nurture the talents and potential of all Americans

Reform Elementary and Secondary Education

The Administration's Race to the Top program uses competitive grants to leverage state and local reform in the quality of elementary and secondary education. The Administration's Blueprint for Reform re-envisions the federal role in K-12 education to create the context for innovation, and the Administration is working with Congress to update and improve the Elementary and Secondary Education Act (as amended by the No Child Left Behind Act). The President's FY 2012 Budget includes support for School Turnaround Grants to drive change and improve student performance in 5,000 of the nation's lowest-performing schools, and support for Investing in Innovation to develop and expand effective innovative strategies to improve students' educational outcomes.

Restore America to First in the World in College Attainment

America has slipped from 1st to 9th among OECD countries in the share of young people with a college degree. President Obama is committed to restoring America to first place by 2020. The Health Care and Education Reconciliation Act (HCERA), signed in March 2010, makes all federal loans available directly to students, ending wasteful subsidies once paid to third-party administrators. By saving $68 billion in subsidies over the next 11 years, the direct loan program allows for both deficit reduction and investments in college affordability for low-income students, including a $40 billion expansion of the Pell Grant program. The President will continue efforts to make college affordable and is calling on Congress to make permanent his American Opportunity Tax Credit, worth $10,000 for four years of college. The Trade Adjustment Act invests in our nation's community colleges, enhancing this important avenue to advanced training, which is also a critical component of the educational pipeline for many underserved students who wish to pursue further studies. Finally, the Task Force on Skills for America's Future will build and improve partnerships between businesses and educational institutions to train American workers for 21st century jobs.

Create a First-Class System of Early Education

According to Nobel-award winning economist James Heckman, "Early childhood education fosters cognitive skills along with attentiveness, motivation, self-control and sociability—the character skills that turn knowledge into know-how and people into productive citizens." To achieve these objectives, the Administration is investing in evidence-based early learning and development, and working to hold Head Start programs more accountable by measuring classroom quality and requiring lower-performing programs to compete for funding. In addition, the proposed Early Learning Challenge Fund will upgrade state innovation in early childhood programs through high standards and professional development.

Strengthen and Broaden American Leadership in Fundamental Research

America's universities, federal labs, and industrial laboratories must continue to do the research that will lead to breakthrough products and new companies. That is why the Administration has pushed for historic increases in America's R&D investments. Taken together, these steps will ensure that America continues to generate the most valuable ideas and that Americans have the know-how to implement these ideas here at home.

Enact the Largest R&D Increase in Our Nation's History

With $18.3 billion in research funding, the Recovery Act was part of the largest annual increase in research and development in America's history. The President's FY 2012 Budget provides additional support for science and basic research, delivering on the President's commitment to double funding for three key basic research agencies—the National Science Foundation, the Department of Energy's Office of Science, and the National Institute of Standards and Technology laboratories.

Set National Goal to Invest Three Percent of GDP in R&D

The President has set a goal for America to invest more than three percent of our GDP in public and private research and development. This investment rate will surpass the level achieved at the height of the space race, and can be achieved through policies that support basic and applied research, create new incentives for private innovation, promote breakthroughs in national priority areas, and improve STEM education.

Build a Leading Physical Infrastructure

In the 20th century, the United States built highway and aviation networks that transformed the country—connecting our markets, businesses, and workers to fuel rapid economic expansion. Today, new transportation investments are needed to support economic growth and keep America competitive in the global economy. Efficient transportation systems facilitate trade and collaboration, and help workers find well-matched but distant jobs. By integrating our markets, infrastructure investments create the scale that attracts innovation and the competition that causes best practices to spread.

Fulfill a New Transportation Vision with High-Speed Rail

President Obama has set the ambitious goal of connecting 80% of Americans to the high-speed rail system within 25 years. To accomplish this, he proposes sustained investments that build on the Recovery Act and would help create an efficient, high-speed passenger rail network of 100- to 600-mile intercity corridors that will better connect communities across America. This vision builds on the successful traditional highway and aviation development models with a 21st century solution that focuses on a clean, energy-efficient option.

Develop the Next Generation of Air Traffic Control

The Next Generation Air Transportation System (NextGen) is a game-changing investment in our national airspace system to make air travel more convenient, dependable, and energy-efficient, while ensuring flights are as safe and secure as possible. President Obama's FY 2012 Budget increases funding for this critical innovation.

Create a National Infrastructure Bank

The President has proposed the creation of a National Infrastructure Bank, which will provide a new way to leverage investments in the nation's highest priority infrastructure projects, supplementing reformed formula-grant transportation programs.

Develop an Advanced Information Technology Ecosystem

Just as our physical infrastructure provided a foundation for decades of economic growth, so will America's information technology ecosystem catalyze growth for decades to come. By investing in broadband and a modernized electric grid, securing cyberspace and ensuring the efficient use of our wireless spectrum, the U.S. is laying the groundwork for sustainable economic growth, well-paying jobs, and global competitiveness, while enabling innovations that today cannot even be imagined.

Develop a Nationwide, State-of-the-Art Communication Network

The Administration is committed to facilitating the development and deployment of a next-generation wireless broadband network that can reach at least 98% of Americans and enable public safety to have access to a nationwide and interoperable wireless network. A key component of catalyzing investment and innovation in state-of-the-art wireless technology is freeing up more spectrum for wireless broadband. To that end, the Administration is sharply increasing, from 50 MHz to 550 MHz, the amount of wireless spectrum that will be made available for commercial use. The Administration's ten-year plan will avoid "spectrum crunch" and facilitate the ongoing torrent of innovation seen in smartphones, netbooks and tablets, and the applications that run on them.

Expand Access to Broadband

Approximately 100 million Americans still do not subscribe to broadband at home, leaving one-third of our households operating with 20th century infrastructure in a 21st century economy. The Recovery Act provided $6.9 billion to substantially expand broadband access for households, businesses, schools, libraries, public safety providers, and hospitals across America.

Modernize the Electric Grid

By setting standards for smart grid technologies and making information technology investments, the Administration is bringing the nation's electricity grid into the 21st century to reduce energy waste. In particular, federal investments and policy leadership in this area serve to help consumers and utilities optimize the timing and sourcing of electricity use, which promises to reduce costs, increase reliability, and limit blackouts, while improving the security of the electricity system and enabling it to better use clean energy technologies.

Secure Cyberspace

The National Security Council and Homeland Security Council have provided President Obama with a strategic framework to enhance cyber security. Under the leadership of the White House Cyber Security Coordinator, the federal government is working to secure our information infrastructure through new protocols, improved detection capabilities, and game-changing research and development.

Support Research for Next-Generation Information and Communications Technology

The Obama Administration is dedicated to keeping the U.S. on the cutting edge of IT developments. In addition to the Administration's support for wireless innovation, the Networking and Information Technology Research and Development (NITRD) Program funds research in areas such as high-speed networks, next-generation supercomputers, cyber-physical systems, software engineering, and information management. Recently, the President's Council of Advisors on Science and Technology has identified research directions that will help foster the next revolution in information technology, and transform health care, energy efficiency, education, and transportation.

PROMOTE MARKET-BASED INNOVATION

"All these investments—in innovation, education, and infrastructure— will make America a better place to do business and create jobs. But to help our companies compete, we also have to knock down barriers that stand in the way of their success."

—President Barack Obama, State of the Union Address, January 25, 2011

The private sector is an engine of innovation. Great ideas can come from many corners, and the capacity of decentralized, competitive markets to see innovative opportunities, prove their value, and enable their diffusion drives our economic growth and the creation of new, better jobs for the American people. The Obama Administration is committed to providing the best possible environment for private-sector innovation, whether by established firms or entrepreneurs.

Accelerate Business Innovation with the R&E Tax Credit

The President has proposed to reform the way that businesses are taxed to encourage innovation and entrepreneurship more broadly. Any process or proposal for tax reform should address the market failure that the private incentive for an innovative investment typically falls short of the social interest because many of the innovative benefits accrue to consumers.

Make the R&E Tax Credit Permanent

To encourage private sector innovation, the administration has proposed making the Research and Experimentation Tax Credit permanent, while simplifying its use and expanding its incentive payments by 20%. The proposal for an expanded credit will invest about $100 billion over 10 years in the form of foregone tax receipts to leverage additional innovation.

Promote Ingenuity through Effective Intellectual Property Policy

Intellectual property (IP) rights provide critical incentives for commercial innovation. IP further allows new ideas to be traded between firms, finding their best uses in the marketplace, and is an important determinant of entrepreneurial funding. Because IP supports both innovation and entrepreneurship, public policy must ensure that innovators receive high-quality IP rights in a timely manner, while maintaining public access to basic discoveries and room for healthy experimentation.

Support and Protect Effective Intellectual Property Rights

Our patent system faces serious challenges, with businesses and entrepreneurs now waiting an average of 35 months behind a backlog of over 700,000 patent applications. The Obama Administration is supporting comprehensive patent reform to slash the processing time for patent applications, enable applicants to fast-track their most important applications, and allow a post-grant review procedure that can improve patent quality. These initiatives, coupled with the 2010-2015 Strategic Plan of the U.S. Patent and Trademark Office, work to solve core challenges to our patent system and accelerate innovation, entrepreneurship, and economic growth.

Similarly, effective enforcement of intellectual property rights is essential to innovation and economic growth. In June, the U.S. Intellectual Property Enforcement Coordinator (IPEC) released the Administration's first plan to fight intellectual property infringement. The plan included 33 specific action items, spanning six broad categories, which the Administration agreed to undertake to improve enforcement. The Administration will continue to prioritize intellectual property enforcement and to support U.S. businesses and consumers through protection of intellectual property.

To promote international coordination, the United States Trade Representative has negotiated the Anti-Counterfeiting Trade Agreement, concluded in November 2010, which requires both a strong legal framework for the enforcement against counterfeit and pirated goods and promotes key practices that make those laws effective in reality.

> "Now this is important because small businesses produce most of the new jobs in this country. They are the anchors of our Main Streets. They are part of the promise of America – the idea that if you've got a dream and you're willing to work hard, you can succeed. That's what leads a worker to leave a job to become her own boss. That's what propels a basement inventor to sell a new product – or an amateur chef to open a restaurant. It's this promise that has drawn millions to our shores and made our economy the envy of the world."
>
> — President Obama, at the signing of the Small Business Jobs Act, September 27, 2010

Encourage High-Growth and Innovation-Based Entrepreneurship

Entrepreneurship plays an essential role in generating innovation and stimulating U.S. economic growth. New firms account for most net job growth, and small businesses employ 30% of high-tech workers. Yet market obstacles limit entrepreneurship, as would-be entrepreneurs struggle to raise funding without an established reputation or without giving ideas away. The Obama Administration is committed to helping entrepreneurs build vibrant businesses that lead to new jobs and economic growth.

Increase Access to Capital for New Business

The Small Business Jobs Act (SBJA), signed by President Obama on September 27, 2010, provided an additional $14 billion more in lending support via the Small Business Administration and more than $30 billion in capital support for small business lending via the Treasury, as well as $12 billion in tax relief to small businesses, to help these businesses invest and create jobs. The USDA's Business and Industry Guaranteed Lending Program also provides $1 billion annually and, on account of the Recovery Act, was able to deliver $3 billion in FY 2010 to support the financing of rural businesses. All in all, the Obama Administration has sought to facilitate small business development by cutting taxes on small businesses 17 times.

Hold a Forum on Facilitating Access to Capital for Entrepreneurs

America's preeminence in generating innovative new companies depends on open and well-functioning capital markets. Research indicates that American start-ups create a disproportionate share of new jobs and contribute significantly to economic growth. Recent trends in capital markets, however, present several challenges for American entrepreneurs. To evaluate relevant policy strategies, this forum will bring together top government officials from several agencies with experienced managers and investors from the private sector to assess recent trends in access to capital and its impact on entrepreneurship.

Widen America's Lead as the World's Best Place for High-Growth Entrepreneurship

Startup America is the Administration's sustained campaign to celebrate, inspire, and accelerate high-growth entrepreneurship throughout the nation. This coordinated public/private effort brings together an alliance of the country's most innovative entrepreneurs, corporations, universities, foundations, and other leaders, working in concert with a wide range of federal agencies to dramatically increase the prevalence and success of American entrepreneurs. The core goals of *Startup America* are to increase the number of new high-growth firms that are creating innovation and quality jobs; celebrate and honor entrepreneurship as a core American value and source of competitive advantage; and inspire and empower an ever-greater diversity of communities and individuals to build successful American companies. To achieve these goals, a broad set of federal agencies will launch a coordinated series of policies that ensure high-growth startups have unimpeded access to capital, expanded access to quality mentorship, an improved regulatory environment, and a rapid path to commercialization of federally-funded research.

Promote Regional Innovation Clusters

Regional clusters can be significant sources of entrepreneurship, innovation, and quality jobs, and the root of new industries. The Administration is making substantial investments to promote regional innovation clusters that draw together industry, university, and government resources. The Small Business Administration's Regional Cluster Initiative, the USDA's Agricultural Technology Innovation Partnership Program, and the Department of Energy's Energy Efficient Building Systems Innovation Cluster are all working to spur regional innovation engines in major technology sectors. Moreover, the Economic Development Administration's i6 Challenge series encourages and rewards innovative partnership models that accelerate technology commercialization, new venture formation, and job creation. Finally, the Departments of Labor and Education are aligning Workforce Investment Act training and employment programs and career and technical education with regional innovation clusters to ensure that clusters have the skilled workforce necessary to grow and prosper, and to connect American workers with good career opportunities.

Promote Innovative, Open, and Competitive Markets

Large, efficient markets attract innovative investment, and competition provides the means by which the best ideas spread. Historically, our strong policy against collusion and unfair practices has contributed to vibrant, competitive American businesses that are better equipped to compete in global markets. America's long-standing efforts to open foreign markets have brought our private sector greater innovation incentives and rewards. Well-functioning capital markets have historically made our country a leading place to invest, bringing scarce capital to the best ideas and fueling American ingenuity.

Protect and Enable Competition

The Department of Justice (DOJ) and the Federal Trade Commission (FTC) have developed new Horizontal Merger Guidelines, issued in August 2010. The new Guidelines include, for the first time, a section explaining how the DOJ and FTC assess whether a merger is likely to retard innovation.

Improve Regulation and Regulatory Review

In January 2011, President Obama issued an Executive Order to improve regulation and the regulatory review process. Under this Executive Order, the President required federal agencies to design cost-effective, evidence-based regulations that are compatible with economic growth, job creation, and competitiveness. The Order emphasizes the principles of public participation, integration and innovation, flexible approaches, and scientific integrity. It also includes a provision on retrospective review, which asks agencies to submit a preliminary plan within 120 days to determine whether any regulations should be modified, streamlined, expanded, or repealed so as to make the agency's regulatory program more effective or less burdensome.

Ensure an Open Internet that Protects Consumers and Enables Innovation

President Obama is strongly committed to an open Internet that protects consumers and innovation. To that same end, the FCC has acted to preserve that openness so that users and innovators are able to compete on the merits and not face anticompetitive barriers imposed by incumbent broadband providers. More generally, the Department of Commerce has outlined a path forward to protect Internet privacy, and the Administration has established an interagency effort to guide and address critical policy issues in the Internet environment, including openness, privacy protections, and cybersecurity concerns.

Promote American Exports

In March 2010, President Obama launched the National Export Initiative (NEI), an ambitious effort to help American businesses that sell their goods and services abroad. By unlocking foreign markets for U.S. goods and services, improving access to credit for U.S. businesses, and undertaking other measures, the NEI seeks to double U.S. exports in five years and support millions of additional jobs.

Catalyze Breakthroughs for National Priorities

> "Half a century ago, when the Soviets beat us into space with the launch of a satellite called Sputnik, we had no idea how we would beat them to the moon. The science wasn't even there yet. NASA didn't exist. But after investing in better research and education, we didn't just surpass the Soviets; we unleashed a wave of innovation that created new industries and millions of new jobs. This is our generation's Sputnik moment."
>
> —President Barack Obama, State of the Union Address, January 25, 2010

In areas of well-defined national importance, public investments can help catalyze advances, leveraging key breakthroughs and U.S. leadership. The 21st century brings several critical areas – including energy, bio- and nanotechnology, space capabilities, health care, and education – where the demand for breakthroughs is clear. The Administration's *Strategy for American Innovation* will harness public mechanisms to help meet our common goals, sparking commercial innovations and American ingenuity as we seek to meet the grand challenges of the next century and add impressive new chapters to the history of American progress.

Unleash a Clean Energy Revolution

For our national security, economy, and environment, it is crucial to develop clean energy technologies. President Obama is committed to U.S. leadership in the energy economy of the future. The President's strategy will meet our energy goals and put the U.S. at the cutting edge of the renewable energy, advanced battery, alternative fuel, and advanced vehicle industries.

Double the Nation's Supply of Renewable Energy by the End of 2012

The Administration is committed to doubling the supply of renewable energy by the end of 2012. Federal tax credits and financing support, including the Section 1603 and Section 48C programs, have leveraged the manufacture and deployment of gigawatts of new renewable energy investments in innovative solar, wind, and geothermal energy technologies. Aided by these incentives, electricity generation from renewables (excluding conventional hydropower) is projected to surpass twice its 2008 level, meeting the Administration's goal.

Spur Innovation through New Energy Standards

The President has set a national goal of generating 80% of the nation's electricity from clean sources by 2035. The proposed Clean Energy Standard will mobilize hundreds of billions of dollars in private investment, spur the deployment of clean energy technologies, and create market demand for new innovations. The Administration is also working to meet the renewable fuel mandate set by Congress, which requires the use of 36 billion gallons of renewable fuel by 2022. The EPA finalized a rule to implement the Renewable Fuel Standard on February 3, 2010, and the Growing America's Fuels strategy focuses on a number of the innovations that will help us achieve that goal

Create Energy Innovation Hubs

Bringing together scientists and innovative thinkers from different disciplines to form highly-integrated research teams can create research breakthroughs on tough problems. The Administration established three Energy Innovation Hubs in FY 2010 to tackle challenges in nuclear energy modeling, energy efficiency in buildings, and the generation of fuel from sunlight. The Administration's FY 2012 Budget calls for doubling the number of Energy Innovation Hubs, from three to six, to tackle additional energy challenges.

Invest in Clean Energy Solutions

The Department of Energy's Advanced Research Projects Agency-Energy (ARPA-E) has awarded nearly $400 million to more than 100 research projects that seek fundamental breakthroughs in energy technologies. The President's FY 2012 Budget proposes to expand support for ARPA-E.

Promote Energy Efficient Industries

The Administration is spurring private sector innovation through new fuel efficiency and greenhouse gas emissions standards, with new efforts to develop standards over the 2017-2025 model years for light vehicles and new standards over medium- and heavy-duty vehicles. As the single largest energy consumer in the U.S. economy, government procurement provides an additional, important mechanism to catalyze demand for innovative energy technologies. In October 2009, President Obama signed an Executive Order that calls on agencies to cut the federal government's fleet petroleum use by 30% by 2020.

Invest in Advanced Vehicle Technology

The President's FY 2012 Budget proposes to make the United States the world's leader in manufacturing and deploying next-generation vehicle technologies, expanding funding for vehicle technologies by almost 90% to nearly $590 million and enhancing existing tax incentives. Recovery Act and prior year investments are already making progress on advanced technology vehicles through research initiatives like an ARPA-E grant to develop a battery that will go 300 miles on a single charge. The FY 2012 Budget will significantly broaden R&D investments in technologies like batteries and electric drives – including an over 30% increase in support for vehicle technology R&D and a new Energy Innovation Hub devoted to improving batteries and energy storage for vehicles and beyond. In addition, the President is proposing to transform the existing $7,500 tax credit for electric vehicles into a rebate that will be available to all consumers immediately at the point of sale.

Accelerate Biotechnology, Nanotechnology, and Advanced Manufacturing

The emerging fields of biotechnology and nanotechnology may provide major new platforms for economic growth in the years and decades ahead while providing potentially profound advances in health. The Administration is committed to nurturing these promising fields.

Complete DNA Sequencing for Major Diseases and Drive Sequencing Innovations

Administration investments will lend a new level of insight into disease that was previously unattainable, while driving down the cost of DNA sequencing. As an extraordinary example of the promise afforded by the Recovery Act, the Administration is investing in the sequencing of over 1,800 complete genomes, a more than 50-fold increase over the 34 genomes that have been sequenced to date by non-Recovery Act NIH funding. This effort has the potential to provide critical insight into major diseases while also driving down sequencing costs. The Administration, through NIH, is also leading The Cancer Genome Atlas (TCGA), the largest and most comprehensive analysis of the molecular basis of cancer ever undertaken, which may unleash new possibilities for cancer treatment, diagnosis and personalized care.

Create Nanotechnology Solutions

The National Nanotechnology Initiative is investing in particularly promising nanotechnology areas. Advances in nanotechnology can drive economic growth, create quality jobs, and address a broad range of national challenges. Examples of potential nanotechnology applications include smart anti-cancer therapeutics that target tumors without the devastating side effects of chemotherapy, solar cells as cheap as paint, and the next revolution in computing. Building on basic science advances, companies are already offering nanotechnology-enabled products with breakthrough capabilities in areas such as disease detection, lighter and stronger materials, and next-generation batteries.

Launch Breakthrough Technologies for Advanced Manufacturing

The Administration is investing in breakthrough research for manufacturing technologies that can provide the foundation for future economic growth and competitiveness. The FY 2012 Budget increases investments at key science agencies, including the NSF, NIST laboratories, the Department of Energy Office of Science, and DARPA, to support U.S. leadership in developing advanced manufacturing technologies. The FY 2012 Budget also proposes to initiate the Advanced Manufacturing Technology Consortia program, a public-private partnership that will improve manufacturing R&D investments and accelerate innovations' time to market.

Box 3: The Growing Fields of Nanotechnology and Personalized Medicine

A nanometer is a billionth of a meter – or one hundred thousand times smaller than the diameter of a human hair. Nanotechnology promises to transform multiple industries: capturing and storing clean energy, developing next-generation computer chips, allowing early detection of diseases, creating smart anti-cancer therapeutics that deliver drugs only to tumor cells, and enabling all-new approaches to a wide range of manufacturing activities, among many other examples. While the commercial impact of nanotechnology to date has been limited primarily to nanomaterials applied to a range of consumer goods from health care and food products to textiles, automotive composites and industrial coatings, nanotechnology innovation is beginning to accelerate. The ten-year history of U.S. leadership in fundamental nanotechnology research and development under the National Nanotechnology Initiative has laid the crucial groundwork for developing commercial applications and scaling up production, creating demand for many new nanotechnology and manufacturing jobs in the near-term.

Nanotechnology is being applied in the developing medical engineering and personalized medicine industries. The practice of tailoring medical treatment to an individual's unique genetic make-up makes not only treatment, but early detection and prevention, more effective. It also reduces medical costs in cases where expensive treatments are unnecessary or futile. Researchers are currently experimenting with nanotechnology to develop drugs capable of targeting a disease without triggering the body's natural immune response.

Develop Breakthrough Space Capabilities and Applications

Space technologies warn us of natural disasters, enable worldwide communications, facilitate global commerce, enhance our global navigation and position location capabilities, and contribute to our national security. The President is committed to continuing U.S. leadership in this arena by strengthening our space-related science and technology R&D initiatives, enhancing mutually beneficial international cooperation efforts, improving the state of the U.S. space industrial base, and taking steps to enable a robust, innovative, and competitive commercial space sector.

The Administration is dedicated to developing the next generation of space vehicles and innovative uses of the International Space Station (ISS) by working with the private sector to expand American industry's role. The Administration is also committed to advancing U.S. capabilities in other space sectors, including a new generation of global positioning satellites and services that allow advanced navigation and timing applications, which can be a platform for further innovation in many sectors, including agriculture, communications, air travel, and highway safety. The Administration will also continue and improve a broad array of programs of space-based observation, research, and analysis of the Earth's land, oceans, and atmosphere.

Drive Breakthroughs in Health Care Technology

The inefficiencies in our health care system raise costs and reduce the quality of care. Information technology has the potential to revolutionize our health care systems, creating technological platforms to reduce costs, reduce errors, and increase the quality of care. By breaking down barriers between health service providers, health IT can integrate our health markets to attract scalable, competitively tested private sector innovation.

Expand the Use of Health IT

The Office of the National Coordinator (ONC) for Health Information Technology is promoting health IT adoption through numerous initiatives. These efforts include programs to accelerate adoption of Electronic Health Records, to develop standards for health information exchange over the Internet, and to develop mobile health technologies. In addition, the Strategic Health IT Advanced Research Projects (SHARP) Program funds potentially game-changing advances to address well-documented problems that have impeded adoption of health IT.

Enable Innovation in Medical Technologies and Medical Care

The U.S. Food and Drug Administration is working to accelerate the development of medical device technologies. In June 2009, the FDA launched a "Transparency Initiative" which will improve the market's understanding of the approval process and encourage innovation. In January 2010, the FDA announced the creation of the Council on Medical Device Innovation, designed to encourage innovations that will address unmet public health needs. The multi-agency council will facilitate medical technology innovations by identifying the most important unmet public health needs, delineating the barriers to the development of technologies to treat these needs, and addressing obstacles to patient access.

Over time, the Affordable Care Act will reduce costs throughout the health care system by reducing waste, limiting overpayments, and improving incentives for physicians, hospitals, and other providers to focus on the quality of care rather than the volume and price of care. These cost containment strategies will provide strong incentive for process and technology innovations that improve the delivery of health services and lower costs. Projections by the Centers for Medicare and Medicaid Services find that the law will reduce the annualized growth rate of Medicare spending between 2012 and 2019 by 1.4 percentage points.

Create a Quantum Leap in Educational Technologies

To maintain America's economic competitiveness, we must dramatically improve our education system in a relatively short period. This challenge is exacerbated by the resource limitations imposed by the current economic environment and the historically persistent cost increases of the education system. Innovation must be a core building block of our nation's drive to implement new education strategies and systems.

Advance Development of Educational Technologies

There is growing evidence that technology can improve learning outcomes and reduce the time needed for students to gain new skills. With relatively widespread access to broadband, cloud computing, digital devices, and software, the technological conditions are increasingly ripe for the development of advanced educational technologies. The Administration is working to promote educational technology innovations through the Department of Education's National Educational Technology Plan, the National Science Foundation's "Cyberlearning Transforming Education" initiative, and various DARPA, Department of Energy, NOAA, and U.S. Navy programs. To catalyze the development and commercialization of transformative educational technologies, the Administration is supporting the launch of an ARPA-ED in FY 2012, building on successful models used by the DOD and DOE.

Appendix A: Invest in the Building Blocks of American Innovation

Educate Americans with 21st Century Skills and Create a World-Class Workforce

Improve America's Science, Technology, Engineering, and Math (STEM) Education

The President is determined to improve the participation and performance of America's students in science, technology, engineering, and mathematics. To that end, the President has pledged to prepare an additional 100,000 STEM teachers by the end of the decade, with strong teaching skills and deep content knowledge. Moreover, as part of this commitment, President Obama launched the Educate to Innovate campaign on November 23, 2009 (*See Box 4*). The goals of the Educate to Innovate campaign are three-fold:

- Increase STEM literacy so that all students can learn deeply and think critically in science, math, engineering, and technology.
- Move American students from the middle of the pack to the top in the next decade.
- Expand STEM education and career opportunities for underrepresented groups, including women and girls.

Box 4: The Government as a Partner in Educate to Innovate

One of the key roles of government in promoting innovation is to convene the relevant stakeholders and catalyze partnerships to achieve common goals. President Obama's Educate to Innovate campaign is an example of this kind of partnership, where the Administration, leading private sector firms, foundations, and nonprofits are working together to promote science and math education across America.

Recent trends indicate that despite many high quality schools, effective teachers, and successful students, American schoolchildren are falling behind in math and science. In the 2009 Programme for International Student Assessment (PISA), American students placed 17th out of 34 in science and 25th out of 34 in math among developed nations. Our own 2009 National Assessment of Educational Progress (NAEP) found that American 4th graders and 8th graders have shown little or no progress in math scores over the last decade. Educate to Innovate aims to increase STEM literacy and help American students move from the middle to the top of the pack of international performance in STEMIn addition, the campaign aims to broaden STEM education and career opportunities for underrepresented groups.

The President's campaign has already resulted in over $700 million in financial and in-kind support for STEM programs. The President's "all hands on deck" call to improve STEM education has galvanized indus-try, universities, foundations, science and technology centers, libraries, and science and engineering profes-sionals to do more. The private sector is responding not just with financial support, but with commitments that take advantage of their core competencies and the skills and passion of their employees. Over 100 CEOs have come together to launch Change the Equation, a historic effort to scale up effective models for improving STEM education. The President has also personally helped raise the visibility of STEM by holding the first ever White House Science Fair, meeting with students who are developing cancer therapies, water purification systems, and robotic wheelchairs.

Backup Documents: Prepare and inspire: K-12 education in stem for America's future, Supporting STEM Factsheet, Educate to Innovate Homepage

Reform Elementary and Secondary Education

Investment in education must be accompanied by reform and innovation. The Administration is fostering a Race to the Top initiative in our nation's schools, supporting new, state-of-the-art assessment and accountability systems that provide timely and useful information about the progress of individual students. The Administration is also devoted to ensuring that teachers are supported as professionals in the classroom, while also holding them to high professional standards.

- The Race to the Top program uses competitive grants to encourage systematic state and local reform and significantly improve student outcomes. Following its creation with a $4.35 billion allocation in the Recovery Act, the Administration has asked for an additional phase of Race to the Top, in response to the interest demonstrated by states and the meaningful changes states are undertaking. The new competition would place a particular focus on cost-effective reforms that improve student achievement in an era of tight budgets and allow school districts to compete directly for funds.

- The Investing in Innovation program provides grants to school districts and to nonprofit organizations working with school districts to validate or scale up educational strategies for which there is strong evidence of improving educational outcomes, or to develop and test promising practices for which there is potential but whose efficacy has not yet been systematically studied.
- The Administration is working with Congress to reauthorize the Elementary and Secondary Education Act with a new approach. This Act (as most recently reauthorized by the No Child Left Behind Act) is the major federal law regarding K-12 education. The Administration's *Blueprint for Reform* will create a context for innovation by ensuring that states set clear, high standards that meaningfully signify readiness for college and a career.
- The President's FY 2012 Budget includes funding to support School Turnaround Grants, which will support vigorous interventions for 5,000 of the nation's lowest-performing schools over the next five years, designed to drive change, improve student achievement, and transform school culture.

Backup documents: Race to the Top executive summary, Race to the Top legislation and regulations, Phase 1 applications, scores, and comments, Phase 2 applications, Phase 2 finalist press release, President Obama's announcement of the FY11 Budget request for $1.35B–"Speeding up the Race to the Top", Department of Education–Resources About ESEA Reauthorization, Current ESEA language, A Blueprint for Reform, Research Behind the Obama Administration's Proposal for Reauthorizing the Elementary and Secondary Education Act (ESEA). President Obama Announces Steps to Reduce Dropout Rate and Prepare Students for College and Careers, FY2011 Budget Factsheet–Department of Education

Restore America to First in the World in College Attainment

President Obama is committed to ensuring that America will again have the highest proportion of students graduating from college in the world by 2020. To accomplish this goal, the President is committed to increasing higher education access and success by restructuring and expanding college financial aid, by improving opportunities at community colleges, by encouraging partnerships with the public workforce system, and by leveraging business and nonprofit investments in our workforce.

- The Health Care and Education Reconciliation Act (HCERA), signed in March 2010, makes all federal loans—Stafford loans, PLUS loans, and consolidation loans—available directly to students, ending wasteful subsidies once paid to third-party administrators. By saving a projected $68 billion in subsidies over the next 11 years, the direct loan program allows for deficit reduction while also increasing college affordability. In addition to provisions in the Recovery Act, provisions of the HCERA invest more than $40 billion in the Pell Grant program. Pell Grants can be applied toward traditional college expenses as well as to vocational and adult education programs.
- Through the $2 billion Trade Adjustment Assistance Community College and Career Training Grant Program, the Administration is investing in community colleges and other institutions of higher education to help expand and improve their ability to deliver education and career training programs that will prepare participants for employment in high-wage, high-skill occupations. The Program is designed to have a lasting impact on higher education by building on evidence-based strategies, testing innovative approaches, and using data collection and evaluation to build knowledge about which strategies help students succeed. Furthermore, the National Science Foundation is investing in community colleges through its Advanced Technological Education program.
- President Obama has established the Task Force on Skills for America's Future to build better partnerships between businesses, community colleges, and other training providers to get Americans trained for the jobs of today and tomorrow.

In bolstering higher education, the Administration also recognizes an important obligation to support educational institutions that provide opportunity to students in the rural heartland of America.

Backup documents: Reforming Student Loans, Paving the Road to Opportunity, Investing in Pell Grants to Make College Affordable, Building American Skills Through Community Colleges, Presidential Memorandum–Task Force on Skills for America's Future, President Obama to Announce Launch of Skills for America's Future, Skills for America's Future

Create a First-Class System of Early Education

Quality early education is an important component in closing achievement gaps and helping every child reach his or her full potential. The Administration has an ambitious agenda for raising quality by measuring performance, setting standards, and spurring innovation.

- The Child Care and Development Block Grant provides funding to states for child care subsidy programs. Under the Administration's reform principles, states would remain the key policy-makers for child care programs but would be encouraged to set higher standards for child care programs and establish quality rating improvement systems so that parents can make informed choices and providers have incentives to improve.
- Head Start provides early education to nearly one million low-income preschoolers, infants, and toddlers. For Head Start to fulfill its mission of providing low-income children the "head start" they need to succeed in school, lower-performing grantees either need to improve their performance or be replaced with more successful grantees. The Administration is moving forward with an ambitious plan to better measure classroom quality in Head Start Programs and inject competition into the program by requiring lower quality programs to compete for continued funding.
- States have been engaged in K-12 innovation for years, but less attention has been paid at the state level to examining early education and finding ways to set common standards and measure performance across programs. The Early Learning Challenge Fund is meant to foster reform in early education and develop models for early education that can be replicated across the country.

Backup documents: Head Start Roadmap to Excellence, HHS Announces Centers for Excellence, Early Learning Challenge Fund, Department of Education's Early Learning Homepage

Strengthen and Broaden American Leadership in Fundamental Research

Enact the Largest R&D Increase in Our Nation's History

The Administration has made the largest annual increase in research and development in American history, drawing on $18.3 billion in research funding from the Recovery Act. The President's FY 2012 Budget continues to emphasize investments in basic research and delivers on the President's commitment to double funding for three key basic research agencies—the National Science Foundation, the Department of Energy's Office of Science, and the National Institute of Standards and Technology laboratories. These investments will expand the frontiers of human knowledge and create the foundation for our future industries and jobs, helping establish U.S. leadership in new bio-, info-, and nanotechnologies in addition to specific areas such as robotics and "materials by design."

Backup documents: *American Reinvestment and Recovery Act of 2009, Budget Factsheet–Investing in Innovation to Create the Industries and Jobs of Tomorrow*

Set National Goal to Invest Three Percent of GDP in R&D

The President has set a goal for America to invest more than three percent of our GDP in public and private research and development. This investment rate will surpass the level achieved at the height of the space race, and can be achieved through policies that support basic and applied research, create new incentives for private innovation, promote breakthroughs in national priority areas, and improve STEM education.

Backup documents: *Remarks by the President at the National Academy of Sciences Annual Meeting, Remarks by the President on the Economy at Carnegie Mellon University*

Build a Leading Physical Infrastructure

Fulfill a New Transportation Vision with High-Speed Rail

President Obama has set the ambitious goal of connecting 80% of Americans to the high-speed rail system within 25 years. To accomplish this he proposes sustained investments that build on the Recovery Act and would help create an efficient, high-speed passenger rail network of 100- to 600-mile intercity corridors that will better connect communities across America. This vision builds on the successful traditional highway and aviation development models with a 21st century solution that focuses on a clean, energy-efficient option.

Backup documents: *President Obama, Vice President Biden Announce $8 Billion for High-Speed Rail Projects across the Country, Vision for High Speed Rail in America*

Develop the Next Generation of Air Traffic Control

The Next Generation Air Transportation System (NextGen) is a comprehensive overhaul of our national airspace system. NextGen will make air travel more convenient and dependable for passengers, significantly reducing flight delays and increasing airport capacity, while making flights more energy-efficient, safe, and secure. Recognizing the importance of NextGen for our economic prosperity, our environment, and our national security, President Obama's FY 2012 Budget increases funding for this critical innovation.

Backup documents: NextGen Homepage

Create a National Infrastructure Bank

The President has proposed the creation of a National Infrastructure Bank, which will provide a new way to leverage investments in the nation's highest priority infrastructure projects, supplementing reformed formula-grant transportation programs. The proposal marks an important complement to traditional federal investment in infrastructure, as the Bank will base its investment decisions on clear analytical measures, selecting those projects with the greatest return for American taxpayers and leveraging private, state, and local dollars to complete projects as efficiently as possible. The Bank will also promote multi-modal projects, which currently face significant obstacles and bureaucratic delays due to the narrow focus of existing programs that fund specific infrastructure modes.

Backup documents: *President Obama Announces Plan to Renew and Expand America's Roadways*

Develop an Advanced Information Technology Ecosystem

> "If transportation infrastructure was and remains a key source of competitive advantage in the industrial economy, digital infrastructure will be a key source of competitive advantage in the knowledge economy. There is no policy step more important for the digital infrastructure than assuring that scarce spectrum is efficiently allocated."
>
> —Lawrence Summers, Director, National Economic Council, June 28, 2010

Develop a Nationwide, State-of-the-Art Communication Network

The wireless revolution provides great promise for America's future economic prosperity, and the President has announced a plan to help businesses extend the next generation of wireless services to at least 98% of Americans. This Wireless Initiative will enable businesses to grow faster, students to learn more, and public safety officials to access secure, nationwide, and interoperable mobile communications.

America's ability to maintain its global technology leadership hinges critically on the availability of spectrum, as it is the currency of the wireless broadband revolution. Hints of what this revolution has to offer – smartphones, netbooks, and the applications that run on them – are beginning to make their way into American consumer consciousness. However, this promise is threatened by a "spectrum crunch," a crowding of the airwaves used to transfer information wirelessly. Experts believe that the United States will require hundreds of MHz of spectrum in coming years, but we only have 50 MHz in the pipeline for commercial use.

Globally, countries are moving aggressively to promote innovation in the wireless field; the United States must develop our digital infrastructure if we hope to maintain our edge in the ecosystems of networks, technologies, products and applications. President Obama catalyzed this effort by signing a Presidential Memorandum committing the federal government to make available 500 MHz of federal and commercial spectrum over the next 10 years to foster investment and economic growth, and to help create jobs in these new, high-value industries. NTIA released, in November of 2010, a ten-year plan to free up this spectrum and also issued its fast-track report that identified 115 MHz that could be repurposed for wireless broadband.

Backup documents: *Presidential Memorandum: Unleashing the Wireless Broadband Revolution, Fact Sheet: Doubling the Amount of Commercial Spectrum to Unleash the Innovative Potential of Wireless Broadband, Speech Text–Technological Opportunities, Job Creation, and Economic Growth, Plan and Timetable to Make Available 500 Megahertz of Spectrum for Wireless Broadband, An Assessment of the Near-Term Viability of Accommodating Wireless Broadband Systems, NTIA Fact Sheet on Spectrum Plan and Time Table–Fast Track evaluation*

Expand Access to Broadband

The American broadband ecosystem has advanced rapidly. In 2000, only eight million Americans had broadband at home; by 2009, that number had grown to 200 million.[6] But more needs to be done. Approximately 100 million Americans still do not subscribe to broadband at home.[7] They are operating with 20th century infrastructure in a 21st century economy. The Recovery Act provided $6.9 billion to substantially expand broadband access for households, businesses, schools, public safety providers, and hospitals across rural America (*See Box 5*).

Backup documents: *The Recovery Act–Transforming the American Economy Through Innovation, Vice President Biden Announces Recovery Act Investments in Broadband Projects to Bring Jobs, Economic Opportunity to Communities Nationwide, Recovery Act Investments In Broadband: Leveraging Federal Dollars To Create Jobs And Connect America, List of Recovery Act Broadband Awards, Broadband USA*

Box 5: Government Investment In 21st Century Infrastructure: The Case of Broadband

The government can promote innovation by investing in 21st Century infrastructure. The Recovery Act made key investments in rural broadband to reduce the disparities between urban and rural access to high speed Internet. Through a combination of grants and loans, the Department of Commerce's National Telecommunications and Information Administration (NTIA) and the Department of Agriculture's Rural Utilities Service (RUS) provided over $7 billion to fund broadband projects across the nation. NTIA funded "middle mile" projects in underserved or unserved areas, many of which are rural, and focused on expanding public computer center capacity and spurring adoption of broadband service. RUS focused on projects that promote rural economic development and job creation beyond the construction and operation of the facilities themselves.

These investments will connect Americans to emerging opportunities in the service economy, including call centers and software development, which are particularly likely to rely on broadband access. Beyond service industries, broadband access can support existing businesses as they maintain customer relationships, track pricing, and manage inventories more effectively. Broadband connections can help small business owners – especially home-based businesses, which are more common in rural areas – reach new markets. Finally, rural broadband has especially important benefits for online education and training programs and health care delivery.

Modernize the Electric Grid

The Administration is currently taking steps to help modernize our nation's electric grid. For example, the Department of Energy is accelerating the development and deployment of advanced electric grid and digital communications technologies, and grid-scale energy storage projects, through the Smart Grid Investment Grant and Demonstration programs with $4.2 billion in Recovery Act Funds. As part of this grid modernization effort, the National Institute of Standards and Technology (NIST) is leading a public-private initiative to develop a framework and roadmap to develop smart grid interoperability standards. The Recovery Act funding is being leveraged with more than $5.5 billion in private sector funds to deploy a wide spectrum of advanced grid technologies across the nation's transmission and distribution system and into consumer premises. For example, to improve system reliability, several of the Recovery Act smart grid projects involve the installation of more than 800 transmission system sensors, or "Phasor Measurement Units," that can alert grid operators and help prevent minor disturbances from cascading into large outages. Recovery Act funds will help install enough of these sensors to provide near 100% coverage in parts of the country's bulk power transmission system (See Figure 1). To oversee these initiatives and others, the Administration is collaborating with state public utility commissions, consumer groups, electric utilities, equipment manufacturers, and industry leaders to advance and develop a more cost-effective, flexible, and secure national electricity system.

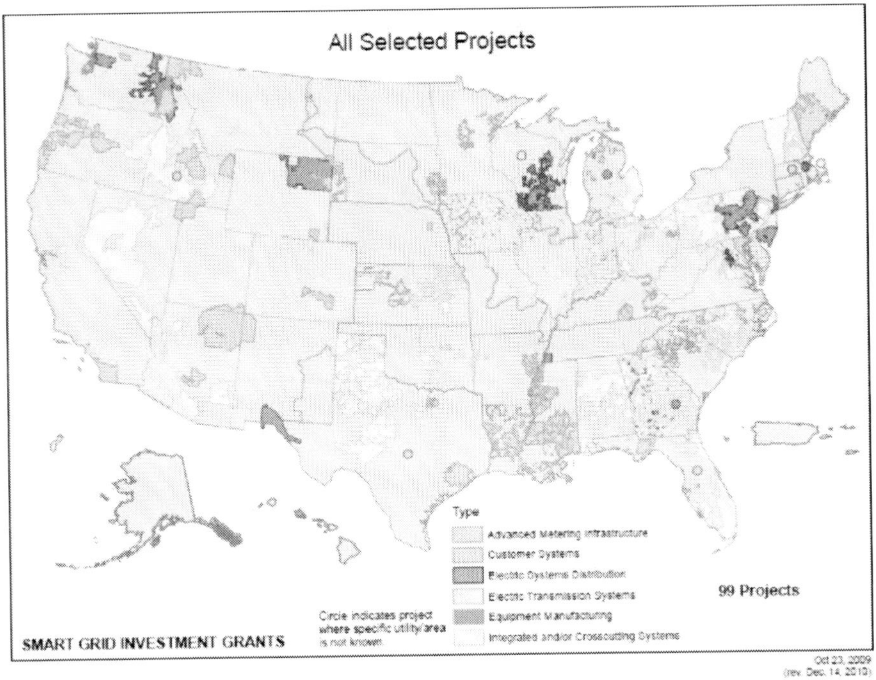

Figure 1.

Backup documents: Smart Grid Homepage, NIST Smart Grid Homepage, NIST Framework and Roadmap for Smart Grid Interoperability Standards–Release 1.0, FERC–National Action Plan on Demand Response, National Science and Technology Council Establishes Subcommittee on Smart Grid

Secure Cyberspace

The National Security Council and Homeland Security Council have provided President Obama with a strategic framework to enhance cyber security. Under the leadership of the White House Cyber Security Coordinator, the federal government is working to secure our information infrastructure through new protocols, improved detection capabilities, and game-changing research and development, including:

- **A framework for game-changing research and development** focused on three key themes: moving-target systems, security tailored to the needs of a particular transaction, and incentives that reward good cybersecurity.
- **Better detection of vulnerabilities and management of security systems using automated and constantly updated tools** that offer enterprises of all sizes the ability to better update security compliance efforts at potentially lower costs and pave the way for future automated protocols.
- **The Administration is leading the way in better securing the Internet's Domain Name System (DNS).** The United States, working with the Internet technical community, deployed a new security technology–domain name system security (DNSSEC)–in the authoritative DNS root. GSA and DHS have also worked to ensure that the ".gov" domain adopts better security practices, including deployment of DNSSEC. In addition, NTIA approved the implementation of DNSSEC within the U.S. top-level domain (.us) and the domain used by academic institutions (.edu).

Backup documents: Cyberspace Policy Review–Final Report, Fact Sheet: Cybersecurity Progress after President Obama's Address, Networking Information Technology Research and Development (NITRD) Game-Change R&D Recommendations, Security Content Automation Protocol (SCAP) Validated Products, NTIA's DNSSEC Page, National Initiative for Cybersecurity Education (NICE)

Support Research for Next-Generation Information and Communications Technology

In addition to specific projects on broadband, smart grid and cybersecurity, the Administration is supporting basic research that will foster future revolutions in information technology. These investments are coordinated through the multi-agency Networking and Information Technology Research and Development (NITRD) Program, which supports research in areas such as high-speed networks, next-generation supercomputers, cyber-physical systems, software engineering, and information management.

A recent review of the NITRD program by the President's Council of Advisors on Science and Technology has identified exciting research directions that will advance information technology and address critical national challenges in health, energy, education and transportation. The President's FY 2012 Budget addresses these recommendations with support for research that will dramatically improve our ability to collect, store, manage and analyze the huge volumes of data that many science and engineering disciplines are now generating.

Backup documents: *NITRD Homepage, PCAST Report*

APPENDIX B: PROMOTE MARKET-BASED INNOVATION

Accelerate Business Innovation with the R&E Tax Credit

Make the R&E Tax Credit Permanent

Expanding, simplifying, and permanently extending the Research and Experimentation tax credit will help companies create good jobs in America now while increasing future productivity and growth. It will also encourage companies to invest in technological development here in America in ways that, while not captured on a company's bottom line, will contribute to our country's innovative capacity. This is a win-win—encouraging job growth and investment now that will pay off with stronger economic growth in the future. Specifically, the President's proposal would:

- **Expand the R&E tax credit by about 20 percent.** This would be the largest increase in the credit in its history. In total, the expanded credit would devote about $100 billion over the next 10 years to leverage additional R&D investment.
- **Simplify the credit.** The Administration proposes to increase the rate of the simpler credit to 17%, which would make it more attractive and simplify tax filing for businesses. Simplifying the credit in this

manner will increase its salience and impact on encouraging investment in research in the United States.

- **Make the credit permanent.** The President would make the credit permanent so that businesses could make investments, create jobs today, and be confident that they will continue to benefit from the credit in the future.

***Backup documents**: R&E tax credit proposal–factsheet*

Promote Ingenuity through Effective Intellectual Property Policy

Support and Protect Effective Intellectual Property Rights

- Optimize Patent Quality and Timeliness | Department of Commerce and USPTO

Timely prosecuted, high-quality patents drive innovation and protect creativity By contrast, delay, uncertainty, poor quality, and costly litigation are obstacles to innovation and creativity. Under the leadership of United States Patent and Trademark Office (USPTO), the Obama Administration has embarked on a multi-year initiative to enhance quality for all and allow speedier examination for those patent applicants who need it, in order to increase the value of the country's entire IP system.

In the past year, USPTO has taken several significant initial steps. For instance, it revised the patent examiner production reward system to increase efficiency, decrease re-work, and support high-quality search and examination. In addition, the USPTO has adopted new, more comprehensive procedures for measuring the quality of patent examination, beginning a process to reengineer the entire patent examination process from the time an application is filed to the patent decision. These efforts coincide with the redesign of the agency's IT infrastructure, which, when complete, will support state-of-the-art work-flow capabilities. The agency also plans to hire more than 1,000 new examiners in the coming two fiscal years in order to increase production capacity.

There are further time-saving and quality-enhancing strategies in development that require necessary public investments. The Obama Administration is supporting comprehensive patent reform legislation coupled with the development of a sustainable funding model for the USPTO.

Innovative programs like the Enhanced Examination Timing Control Initiative ("Three Track") will provide applicants greater control over when their applications are examined and enhance work sharing among IP offices. USPTO administrative actions and Administration *amicus* brief filings can help refine patent law doctrines to increase patent quality. To that end, the USPTO, for the first time in its history, has adopted guidelines for examiners addressing Section 112 of the Patent Act, which governs the "written description and enablement requirements," to improve patent quality by more clearly defining the subject matter protected by a patent grant. Other strategic priorities are included in the USPTO 2010-1015 Strategic Plan, including an ambitious joint goal set by Commerce Secretary Locke and Director Kappos to reduce the average total pendency for a patent application to 20 months from today's 35 month figure.

Backup documents: *USPTO 2010-2015 Strategic Plan, USPTO Public Roundtable on "Three Track" Patent Examination, USPTO is Hiring Patent Examiners, USPTO Adopts New Patent Examination Quality Measurement Procedures.*

- Intellectual Property Enforcement Coordinator (IPEC) | The White House

The PRO-IP Act created the position of U.S. Intellectual Property Enforcement Coordinator (IPEC), placing it within the Executive Office of the President. President Barack Obama nominated, and the Senate confirmed, Victoria Espinel as the first IPEC in late 2009. On June 22, 2010, the IPEC released the Obama Administration's first Joint Strategic Plan to combat intellectual property infringement (JSP). The JSP was the result of significant input from America's innovators – including more than 1,600 public comments on ways to improve enforcement – as well as the collaborative efforts of the Departments of Agriculture, Commerce, Health and Human Services, Homeland Security, Justice, and State, and of White House Offices, including the Office of Management and Budget and the U.S.Trade Representative.

The JSP included 33 action items designed to improve U.S. Government efforts to help protect America's innovation and, thereby, the exports and jobs that come from that innovation. Those action items fell within six broad categories:

1) The U.S. Government will lead by example, respecting intellectual property rights in our policy and our practices.
2) The U.S. Government will be transparent in our policymaking and enforcement, allowing America's innovators and the public to know what the U.S. Government is doing to help enforce their intellectual property rights, and to have input in those efforts.
3) The U S Government will work to ensure efficiency and coordination, both domestically with our federal law enforcement agencies and state and local law enforcement, and internationally with our embassies.
4) The U.S. Government will help enforce our innovators' rights overseas, ensuring that we are able to effectively work with foreign governments when their citizens are infringing U S intellectual property rights (just as we want to effectively work with foreign governments when our citizens are infringing foreign intellectual property rights).
5) The U.S. Government will focus on securing our supply chain, thereby attempting to limit infringing products from entering the U.S. That includes working with Internet service providers, payment processors, advertisers and others to help combat the use of infringing websites to steal America's innovation. In doing so, we will ensure that our efforts are focused on repeated acts of infringement and protecting legitimate uses of the Internet and principles of free speech and fair process.
6) Finally, we will build a data-driven government. Part of this effort includes launching an economic analysis to measure the economic contributions of intellectual property industries We have also committed to comprehensively review existing intellectual property laws to ensure that they are effectively protecting U.S. innovation and U.S. innovators.

Since issuance of the JSP in June, the Office of the IPEC has been working with federal agencies, congressional staff, and the public to implement that strategic plan. The Administration will continue to prioritize intellectual property enforcement and to support U.S. businesses and consumers through protection of intellectual property. The Office of the IPEC will submit, by February 2011, a report to Congress describing implementation.

Backup Documents: 2010 Joint Strategic Plan on Intellectual Property Enforcement, Office of the U.S. Intellectual Property Enforcement Coordinator Homepage

- U.S. Patent Database Available for Free Online | USPTO

Operating from the President's Open Government Directive, the USPTO has worked to vastly increase the amount of U.S. patent information available at no charge on the Internet. The system is now in full production, including all the current feeds that were previously only available by subscription. Also available for the first time is the Patent Application Information Retrieval (PAIR) data, which is the full "wrapper" for a patent application. PAIR data was previously only available on a rate-limited query-only service.

Backup Documents: USPTO Bulk Downloads: Patents

- Negotiated Anti-Counterfeiting Trade Agreement | USTR

In 2010 the Administration delivered on its commitment to strengthen international enforcement of intellectual property rights by concluding substantive negotiations on the proposed Anti-Counterfeiting Trade Agreement (ACTA). ACTA will step up the fight against global proliferation of commercial-scale counterfeiting and piracy in the 21st century. The agreement includes innovative provisions to deepen international cooperation, promote strong enforcement practices, and ultimately help sustain American jobs in innovative and creative industries. In December 2010, the Office of the U.S. Trade Representative (USTR) published a notice seeking public comments in connection with consideration of U.S. signature of the agreement.

Some of the ACTA's key features include commitments to:

- support and enhance approaches to criminal enforcement through stronger requirements for criminal remedies and through improved seizure and destruction of fake goods, seizure of the equipment and materials used in their manufacture, and criminal proceeds;
- combat Internet piracy through a balanced framework that addresses widespread distribution of pirated copyrighted works and preserves fundamental principles such as freedom of expression, fair process, and privacy;

- provide customs authorities with ability to act against import and export shipments as well as to cooperate on in-transit shipments;
- strengthen civil enforcement provisions dealing with damages, provisional measures, recovery of costs and attorneys' fees, and destruction of infringing goods;
- cooperate among ACTA Parties to assist in enforcement efforts; and
- promote good enforcement practices.

Backup Documents: *Anti-Counterfeiting and Trade Agreement (ACTA)*

Encourage High-Growth and Innovation-Based Entrepreneurship

Increase Access to Capital for New Business

- Small Business Jobs Act | Small Business Association

The Small Business Jobs Act (SBJA), signed by President Obama on September 27, 2010, provided an additional $14 billion more in lending support via the Small Business Administration and more than $30 billion in capital support for small business lending via the Treasury, as well as $12 billion in tax relief to small businesses, to help these businesses invest and create jobs. The USDA's Business and Industry Guaranteed Lending Program also provides $1 billion annually and, on account of the Recovery Act, was able to deliver $3 billion in FY 2010 to support the financing of rural businesses. All in all, the Obama Administration has sought to facilitate small business development by cutting taxes on small businesses 17 times.

Backup documents: *President Obama signs small business jobs act*, Small Business Jobs Act Homepage

- Proposed Zero Taxes on Capital Gains from Small Business Investment | The White House

The Tax Relief, Unemployment Insurance Reauthorization, and Job Creation Act of 2010 provides a 100% exclusion from tax for capital gains realized on the sale of certain small business stock held for more than five years. The amount of gain eligible for the exclusion is limited to the greater of $10 million or ten times the taxpayer's basis in the stock. This provision applies to qualified small business stock issued after December 31, 2010, and before January 1, 2012. The President's Budget proposal for FY 2012 calls for making this provision permanent, thereby supporting an expansion of private sector investment in small businesses.

Backup documents: *State of the Union, January 2010, President Obama signs small business jobs act,* Small Business Jobs Act Homepage

Hold a Forum on Facilitating Access to Capital for Entrepreneurs

America's preeminence in generating innovative new companies depends on open and well-functioning capital markets. Research indicates that American start-ups create a disproportionate share of new jobs and contribute significantly to economic growth. However, recent trends in capital markets present several challenges for American entrepreneurs. First, venture capital deals and financing declined significantly during the financial crisis, making it more difficult for early stage companies to develop new products and expand. Second, the IPO market has recovered slowly, reducing the number of successful exits.

Moreover, venture capital and angel funding has historically been localized in areas such as Silicon Valley and Boston. Innovative programs have been launched to encourage capital formation in underserved areas, an especially critical need during this nascent economic recovery.

The forum will bring together top government officials from several agencies with experienced managers and investors from the private sector to assess recent trends in access to capital and its impact on entrepreneurship.

Topics of discussion will include:

- State of the IPO market and ideas to reduce the costs of going public – The early stage capital gap and recommendations to fill it
- Recent trends in venture capital and implications for entrepreneurship
- Perspective of institutional investors on venture capital and private equity – How to channel capital to emerging companies in underserved areas

- Guaranteed Business Lending | USDA

USDA Rural Development's Business and Industry (B&I) Guaranteed Lending Program directly supports the financing of rural businesses, creating sustainable jobs and advancing economic development throughout rural America. The B&I Guaranteed Lending Program improves access to capital for America's rural businesses by providing lending support in partnership with both national and local community banks. The Recovery Act allowed the USDA to provide nearly $3 billion in lending support that was delivered in FY 2010.

Widen America's Lead as the World's Best Place for High-Growth Entrepreneurship

- Startup America | The White House

Our nation has historically been the place where world-leading businesses and new industries take root. Working in concert with the private sector, the Administration is committed to promoting innovation-based, high-growth entrepreneurship – not only in Silicon Valley, Research Triangle Park, and Boston, but all over America. This *Startup America* initiative will expand access to capital and improve the environment for startups; support entrepreneurship education and mentorship programs that empower more people not just to get a job, but to create jobs; strengthen commercialization of university and federally-funded research and development; and ensure that America continues to draw on the most talented innovators and entrepreneurs from around the world in starting new businesses.

To support *Startup America's* goal of expanded access to capital, the Small Business Administration (SBA) will work with the private sector to invest $2 billion over the next five years in high-growth companies in underserved communities and innovation-based startups, all at no cost to taxpayers. The Treasury Department is also working to reform the New Markets Tax Credit, making it easier to support startups in lower-income communities, while the Administration has proposed expanding the credit in its FY 2012 Budget proposal. The Administration will also launch public online suggestion tools and go on the road to cities across America to hear directly from entrepreneurs about the challenges they face and what the Administration can do to reduce undue regulatory burdens and accelerate success. To improve access to mentoring programs, both the SBA and the Veterans Administration have launched programs designed to provide high-quality mentorship by seasoned, successful entrepreneurs to new companies. To help drive university and federally-funded research into the marketplace, the Department of Commerce is launching its second annual i6 Challenge, which will focus on the green tech sector. Commerce is also announcing a Three Track Examination initiative that will give innovators more control over the patent application processing while supporting a more efficient market for innovation.

These initiatives represent just the beginning of a sustained Administration effort to promote high-growth U.S. entrepreneurship. At the same time, responding to the President's call to action, a new coalition of entrepreneurs and other leaders will develop a series of major private-sector commitments aimed at promoting entrepreneurial success and job creation across the country.

Overall, these efforts will serve the three key strategic goals of *Startup America*:

- Increase the number of start-ups and successful high-growth firms that are creating economic growth, innovation, and quality jobs;
- Celebrate and honor entrepreneurship, and reinforce entrepreneurship as a core American value and source of competitive advantage; and
- Broaden the circle of opportunity, so that more communities and Americans participate in our "innovation nation."

Backup Documents: *Startup America Fact Sheet.*

- Promoting High-Growth Entrepreneurship | The Department of Commerce

The Department of Commerce's Office of Innovation and Entrepreneurship promotes innovation-based, high-growth entrepreneurship in pursuit of job creation and economic growth. The Office plays a leading role in developing policy recommendations and implementing initiatives to increase the efficiency and effectiveness of efforts to commercialize technology developed through university and federally-funded research. It manages the i6 Challenge, a multi-agency competitive grant program that encourages and rewards innovative partnership models, such as proof of concept centers, that accelerate technology commercialization, new venture formation, and job creation. It also manages the National Advisory Council on Innovation and Entrepreneurship, which was established to advise the

Secretary and administration on how best to foster high-growth entrepreneurship and support the development of innovation ecosystems.

Backup Documents: EDA homepage, OIE homepage*, National Advisory Council on Innovation and Entrepreneurship, i6 Challenge*

- Innovation Fund | Small Business Administration

The Innovation Fund will support up to $1 billion in private-sector financing over the next five years by matching private capital raised by investment funds that are seeking to deploy capital in early stage innovative small businesses. Access to these sources of capital is a particular challenge that has been documented by the National Academy of Sciences, the Council on Competitiveness, and others. For the past several years, high growth companies that create the industries of tomorrow have encountered serious financing hurdles, leaving many promising technologies and companies stranded in the so-called "Valley of Death" where they are unable to access sufficient capital to grow. Investment funds licensed under the SBIC program raise private capital, a portion of which is guaranteed by SBA. Over the past decade, SBA has supported investment funds that have invested in 6,600 small businesses.

- Reduced Health-Insurance Barriers to Business Creation | Health and Human Services

The Affordable Care Act will reduce health care costs for individuals and reduce health care constraints on entrepreneurship. The Act allows Americans under the age of 26 to remain on their parents insurance and will prohibit new insurance policies from discriminating against Americans with pre-existing conditions, making it easier for Americans to start or join a new company without facing the concern that they will lack health care coverage.

Backup Documents: *Is Employer-Based Health Insurance a Barrier to Entrepreneurship?, Two New Studies: Health Reform Benefits Small Business, The Effects of the Affordable Care Act on Workers' Health Insurance Coverage, Small Businesses and the Affordable Care Act of 2010*

Promote Regional Innovation Clusters

- Regional Cluster Initiative | Small Business Administration

SBA's Regional Innovation Clusters (RICs) harness the potential of regional clusters to drive economic growth, create jobs, and strengthen American competitiveness. Leveraging regional strengths, the RIC program will utilize SBA resources to achieve maximum impact. At the same time, recognizing the needs for technology and innovation in specific areas, SBA is working closely with the Department of Defense on the Advanced Defense Technology (ADT) RIC program, which targets clusters focused on areas of interest to DOD, such as robotics, energy, lightweight materials, and cyber-security.

Backup documents: SBA Regional Clusters Initiative Homepage, *SBA Announces Funding Available to Support Regional Clusters, Job Creation*

- Regional Innovation Ecosystem Development | Department of Commerce

The Department of Commerce's Economic Development Administration (EDA) is leading initiatives at several levels to support the development of regional innovation clusters as building blocks of job creation and U.S. economic competitiveness. EDA is pursuing RIC initiatives on three fronts: (1) Data, Tools, and Best Practices; (2) 21st Century Infrastructure; and (3) Institutional Support. Working through its traditional investment programs such as Public Works and Economic Adjustment, EDA has provided significant support to RICs at the regional level (100+ grants totaling approximately $65 million in FY 2010 Q4 alone). EDA is also committed to streamlining federal support for RICs across agencies, working with partners through the Taskforce for the Advancement of Regional Innovation Clusters (TARIC) on policy recommendations, coordinated grants, and other initiatives.

Backup documents: EDA homepage, OIE homepage, *Regional Innovation Clusters, Regional Innovation Acceleration Network Press Release, Cluster Mapping Press Release, Know Your Region*

- Rural Regional Innovation Initiative | USDA

USDA's Regional Innovation Initiative is designed to support economic development efforts in rural America that deliver robust regional strategies focused on the region's assets. USDA is particularly interested in supporting regional strategies that take advantage of economic opportunities in rural America, including those involving regional food systems, renewable energy, and new jobs related to broadband and rural recreation The initiative encourages asset-based strategic planning that encourages broad-based collaboration.

Backup documents: *USDA Announces Awards to Support Regional Economic Strategies*

- Agricultural Technology Innovation Partnership Program (ATIP) | USDA

USDA Agricultural Research Service (ARS) established the Agricultural Technology Innovation Partnership Program (ATIP) to provide opportunities for the private sector to commercialize research outcomes arising from the vast R&D programs of USDA. ATIP was established to strategically form nationwide geographic partnerships with well-established economic development entities that excel in providing the complementary business assets (manufacturing, marketing, fiscal) needed by private sector entrepreneurs and small businesses to commercialize USDA innovations.

Promote Innovative, Open, and Competitive Markets

Protect and Enable Competition

- Revised Horizontal Merger Guidelines | Antitrust Division of the Department of Justice and the Federal Trade Commission

In September 2009, the Antitrust Division of the Department of Justice (DOJ) and the Federal Trade Commission (FTC) initiated a process to evaluate and possibly update their jointly issued Horizontal Merger Guidelines. These Guidelines describe the merger enforcement policies of the DOJ and FTC. They have a significant impact on the antitrust evaluation of mergers and acquisitions that are subject to review by the DOJ and FTC. Well-designed merger enforcement policies promote innovation by allowing mergers to go forward that are likely to create efficiencies and spur innovation, while preventing mergers that may retard innovation by substantially reducing competition. Revised Guidelines were released in August 2010 and include, for the first time, a section explaining in general terms how the DOJ and FTC assess whether a merger is likely to retard innovation.

Backup documents: *FTC and DOJ Horizontal Merger Guidelines–August 2010*

- Thoughtful and Effective Antitrust Enforcement | Antitrust Division of the Department of Justice and the Federal Trade Commission

The Obama Administration has made balanced antitrust enforcement a core part of its efforts to protect and promote competitive markets. Where mergers or market restraints threaten to prevent firms from introducing innovative products or services, the antitrust agencies—the DOJ and FTC—are vigilant in addressing such conduct. The Obama Administration also recognizes that American companies often need to collaborate to compete effectively in worldwide markets. The antitrust agencies are advocates at home (with federal and state regulatory agencies) and abroad for the principle that sound competition policy means protecting the competitive process—allowing firms to succeed or fail on the merits—and not individual competitors.

Improve Regulation and Regulatory Review

In January 2011, President Obama issued an Executive Order to improve regulation and the regulatory review process. Under this Executive Order, the President required Federal agencies to design cost-effective, evidence-based regulations that are compatible with economic growth, job creation, and competitiveness. The Order emphasizes the principles of public participation, integration and innovation, flexible approaches, and scientific integrity. It also includes a provision on retrospective review, which asks agencies to submit a preliminary plan within 120 days to determine whether any regulations should be modified, streamlined, expanded, or repealed so as to make the agency's regulatory program more effective or less burdensome.

Ensure an Open Internet that Protects Consumers and Enables Innovation

President Obama is strongly committed to protecting an open Internet that protects consumers and innovation. To that same end, the FCC has acted to ensure that users and innovators are able to compete on the merits and not face anticompetitive barriers posed by incumbent broadband providers. Similarly, President Obama is committed to ensuring that the Internet remains an environment where commerce can prosper and users will trust that they will be treated fairly. To that end, the Department of Commerce has outlined a path forward on protecting Internet privacy. Finally, as to the development of Internet policy principles more generally—protecting privacy, ensuring an open Internet, addressing cybersecurity concerns, etc. —the Administration has established an interagency effort to develop guidance on how to address policy issues in the Internet environment.

Backup Documents: *Commerce Privacy Paper, FCC Action on Open Internet, White House Statement on FCC Action On Open Internet, Formation of NSTC Subcommittee on Privacy and Internet Policy Principles*

Promote American Exports

- National Export Initiative | The White House

In March 2010, President Obama signed an executive order launching the National Export Initiative (NEI), an ambitious effort to help American businesses that sell their goods and services abroad. The NEI aims to fulfill the President's goal of doubling U.S. exports in five years in support of millions of additional jobs. As part of this effort, the Administration is focusing on five chief areas, including: access to credit, especially for small and medium-sized firms; increased trade advocacy and export promotion efforts; removal of barriers to the sale of U.S. goods and services abroad; enforcement of trade rules; and the pursuit of policies that will increase global economic growth so that there's a strong worldwide market for U.S. goods and services.

The NEI Executive Order established an Export Promotion Cabinet, which includes the Secretaries of Commerce, State, Treasury, Agriculture and Labor, the U.S. Trade Representative, and the heads of all the trade-related government agencies. Meeting individually and as part of working groups convened by the Trade Promotion Coordinating Committee (TPCC), the Export Promotion Cabinet and other federal agencies have developed short- and long-term recommendations in the priority areas identified in the NEI Executive Order.

In September 2010, the Export Promotion Cabinet delivered a report to the President providing an overview of the progress of the NEI to date and laying out a comprehensive plan for reaching the President's goals. The NEI Report presents the Export Promotion Cabinet's recommendations for doubling exports in five years. It will be followed by the *National Export Strategy*, which will detail the implementation of these recommendations and measure progress. The National Export Strategy is delivered to Congress annually and is expected in March 2011. Exports in the first 11 months of 2010 were up 17% from a year earlier, above the rate needed to reach the President's goal.

In early December 2010, the United States and South Korea reached agreement on a trade deal in which the tariff reductions alone are expected to boost annual exports of American goods by up to $11 billion. The opening of the Korean goods and services market to U.S. exports is expected to support at least 70,000 American jobs.

Backup Documents: *Executive order–National Export Initiative, NEI homepage, NEI Report to the President (September 2010)*

Appendix C: Catalyze Breakthroughs for National Priorities

Unleash a Clean Energy Revolution

Double the Nation's Supply of Renewable Energy by the End of 2012

- 1603 and 48C Grant Programs | Department of Treasury

Federal tax credit and related programs have spurred deployment of innovative solar, wind, and geothermal energy technologies. The wind industry has led the way, with over nine gigawatts (GW) of new generating capacity installed in 2009 alone, aided by Recovery Act incentives. The Section 1603 Payments-In-Lieu-Of-Tax-Credits program supported more than 200 megawatts (MW) of solar projects and more than 100 wind projects around the country, totaling 5.3 GW of wind power capacity. The Section 48C Advanced Energy Manufacturing Tax Credits, as well as loan guarantees, are helping thin-film solar manufacturers create manufacturing capacity for 230 MW per year. The 1603 and 48C incentives have also supported new geothermal projects, including projects with a capacity to generate 112 MW.

Finally, the U.S. Department of Energy has issued loan guarantees or conditional commitments to support nearly 2 GW of new solar, geothermal, and wind capacity. Aided by these incentives, electricity generation from renewables (excluding conventional hydropower) is projected to be twice its 2008 level, meeting the Administration's goal of doubling the supply of renewable energy in three years (*See Figure 2*).

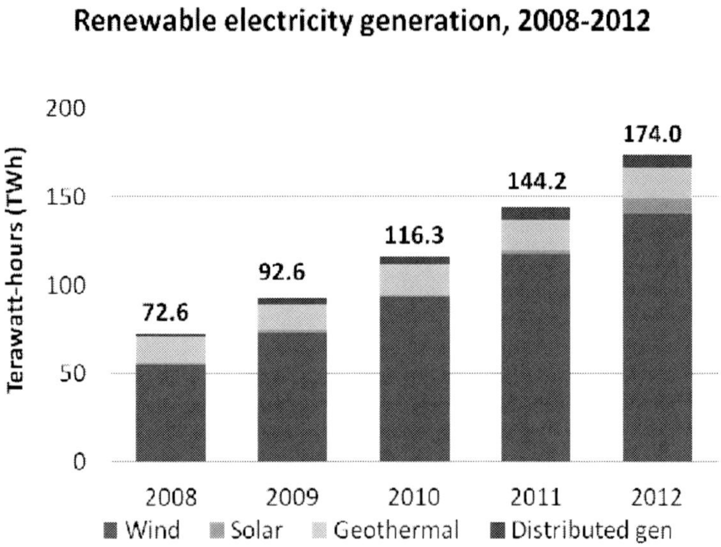

Figure 2. Energy Information Administration data and forecasts.

Backup documents: *Advanced Energy Manufacturing Tax Credit (48C), Fact Sheet: $2.3 Billion in New Clean Energy Manufacturing Tax Credits, Preliminary Evaluation of the Impact of the Section 1603 Treasury Grant Program on Renewable Energy Deployment in 2009, The Recovery Act: Transforming the American Economy through Innovation*

Spur Innovation through New Energy Standards

- Clean Energy Standard | The White House

In the State of the Union Address on January 25, 2011, President Obama announced an initiative to produce more electricity through clean energy, proposing an ambitious but achievable goal of generating 80% of the nation's electricity from clean energy sources by 2035. This target will help position the United States as a global leader in developing and manufacturing cutting-edge clean energy technologies, improving the nation's economic growth and competitiveness while also protecting the environment and developing an energy future that is secure and independent.

As an important step to reach this goal, the President is proposing a new Clean Energy Standard (CES) that will be founded on five core principles: (1) double the share of clean electricity in 25 years; (2) credit a broad range of clean energy sources; (3) protect consumers against rising energy bills; (4) ensure fairness among regions; and (5) promote new technologies such as clean coal.

The CES should work to steadily increase the share of delivered electricity generated from clean energy sources, rising from 40% today to 80% by 2035. To ensure broad deployment and provide maximum flexibility in meeting the target, clean energy credits should be issued for electricity generated from renewable and nuclear power, with partial credits given for clean coal and efficient natural gas. Smart policies will help American families and businesses save money by saving energy. For example, the CES can be paired with energy efficiency programs that will lower consumers' energy bills (e.g. stronger appliance efficiency standards, tax credits for energy efficiency upgrades, and the proposed Home Star program), and will include provisions to help manufacturers invest in technologies to improve efficiency and reduce energy costs. Further, because different regions of the country rely on diverse energy sources today and have varying access to future clean energy resources, the CES must ensure that these differences are taken into account. Finally, the CES should encourage the deployment of emerging technologies, such as clean coal with carbon capture and sequestration.

- Bioenergy | USDA and EPA

On February 3, 2010, the EPA finalized a rule to implement the Renewable Fuel Standard, a long-term renewable fuel mandate established by Congress that requires the use of 36 billion gallons of renewable transportation fuel by 2022. Advanced, low-carbon renewable fuels such as cellulosic biofuel must make up 21 billion gallons of this mandate. Achieving this mandate will help speed the transition to cleaner, more secure sources of energy in the transportation sector, helping our nation address the challenges of climate change, dependence on oil, and job creation. To do so, the Administration is acting on multiple fronts, such as creating the Biomass Crop Assistance

Program and increasing the amount of ethanol that is allowed to be blended in certain cars from 10% to 15% (E15).

With a focus on helping the country reach 36 billion gallons by 2022, USDA, in collaboration with DOE and EPA, developed the *Growing America's Fuels* strategy. This plan will help ensure that dependable supplies of feedstock are available for the production of advanced biofuels to meet legislated goals and market demand, as well to enhance rural economic sustainability. Toward that end, USDA is establishing five Regional Biomass Research Centers and has published a Biofuels Production Roadmap addressing regional variations in feedstock availability and biorefinery locations. In addition, USDA is implementing bioenergy programs from the 2008 Farm Bill, including the Biomass Crop Assistance Program, which provides payments for biomass feedstocks, and Sections 9003, 9004, and 9005, which support producers of advanced biofuels.

Backup documents: *USDA Energy Fact Sheet, USDA Regional Biomass Research Centers, Press Release: Obama Announces Steps to Boost Biofuels, Clean Coal; February 3, 2010, EPA's Renewable Fuel Standard Homepage, Biomass Crop Assistance Program, EPA–E15*

Create Energy Innovation Hubs

- Energy Innovation Hubs | Department of Energy

The urgency of the energy and climate challenges facing our nation, coupled with challenges to our nation's leadership in technological innovation, led to the establishment of the first Energy Innovation Hubs. These Hubs are indicative of the Administration's new, proactive approach to research and a way to harness American ingenuity to drive ground-breaking technology. Energy Innovation Hubs are large, multi-disciplinary, highly-collaborative teams of scientists and engineers working to achieve a specific high priority goal. They are led by top researchers with the knowledge, resources, and authority to nimbly guide efforts, seizing new opportunities or closing off unproductive lines of research. Funding Opportunity Announcements (FOAs) have been issued for three hubs: Energy Regional Innovation Cluster (efficient buildings), Liquid Fuels from Sunlight, and Modeling and Simulation on Nuclear Reactors (nuclear energy).

The first three Hubs were competitively awarded in 2010: the Modeling and Simulation for Nuclear Reactors Hub was awarded to the Consortium for

Advanced Simulation of Light Water Reactors, led by Oak Ridge National Laboratory; the Joint Center for Artificial Photosynthesis, a team co-led by the California Institute of Technology and Lawrence Berkeley National Laboratory, was selected for the Fuels from Sunlight Energy Innovation Hub; and the Greater Philadelphia Innovation Cluster for Energy-Efficient Buildings, a team led by the Pennsylvania State University was chosen to establish the Innovation Hub focused on developing technologies to make buildings more energy efficient. To solve further, key challenges, the President's FY 2012 Budget calls for doubling the number of Energy Innovation Hubs.

Backup documents: *Energy Innovation Hub Homepage, Factsheet: Energy Innovation Hub for Modeling and Simulation, Consortium for Advanced Simulation of Light Water Reactors, Energy Efficient Building Systems Regional Innovation Cluster Initiative, Fuels from Sunlight, Joint Center for Artificial Photosynthesis, Penn State to Lead Philadelphia-Based Team that will Pioneer New Energy-Efficient Building Designs, Greater Philadelphia Innovation Cluster for Energy-Efficient Buildings.*

Invest in Clean Energy Solutions

- ARPA-E Grants for Transformational Technology | Department of Energy

The Department of Energy has supported more than 100 ambitious research projects that could fundamentally change the way the country uses and produces energy. Funded through the Department of Energy's Advanced Research Projects Agency-Energy (ARPA-E), nearly $400 million was awarded to projects that could produce advanced biofuels more efficiently from renewable electricity; design completely new types of batteries to make electric vehicles more affordable; remove carbon pollution from coal-fired power plants in a more cost-effective way; and pursue other game-changing technologies. The Administration's FY 2012 Budget proposes to expand funding for ARPA-E.

Backup Documents: ARPA-E Homepage

- Advanced Technology Vehicles Manufacturing (ATVM) Loan Program | Department of Energy

The President has set an ambitious goal of putting 1 million advanced technology vehicles on the road by 2015 – which would reduce dependence on foreign oil and lead to a reduction in oil consumption of about 750 million barrels through 2030. The Advanced Technology Vehicles Manufacturing (ATVM) Loan Program is investing in automobile manufacturers and component suppliers for projects that reequip, expand, and establish manufacturing facilities in the United States to produce light-duty vehicles and components for such vehicles, which provide meaningful improvements in fuel economy performance. In 2009, the U.S. had only two factories manufacturing advanced vehicle batteries that power advanced technology vehicles and produced less than two percent of the world's advanced batteries. The American Recovery and Reinvestment Act included $2.4 billion for battery and electric drive component manufacturing, and for electric drive demonstration and infrastructure – investments that are already transforming the advanced vehicle batteries industry in the U.S. As a result of these investments, in just the next few years, battery costs are expected to drop by half (2009-2013), the United States will be able to produce enough batteries and components to support 500,000 plug-in and hybrid vehicles and will have the capacity to produce 40% of the world's advanced batteries (2015). The Recovery Act is also supporting the deployment of infrastructure for advanced technology vehicles. And this year's Budget will significantly broaden R&D investments in technologies like batteries and electric drives – including an over 30% increase in support for vehicle technology R&D and a new Energy Innovation Hub devoted to improving batteries and energy storage for vehicles and beyond.

Backup Documents: ATVM Loan Program Homepage, *ATVM Loan Program Factsheet*

- Investments in Advanced Battery and Electric Drive Component Technology | Department of Energy

Electric cars need batteries and drive components that are in short supply in the U.S. To stimulate their production, over $2 billion in Recovery Act funding is being invested in companies like A123 (*See Box 6*) and EnerDel, supporting 30 factories to produce the advanced batteries and electric drive components necessary to power the electric-drive vehicles of the future. By 2015, when these factories reach scale, they will have the capacity to produce enough batteries to support up to 500,000 plug-in hybrid electric vehicles.

From a negligible portion of the world's advanced battery manufacturing today, U.S. production capacity for advanced vehicle batteries will amount to more than 20% of global production capacity estimated to be online in 2012. If successful, these breakthroughs could cut battery costs by as much as 90% and expand range three- to six-fold, bringing the up-front cost of electric cars down to that of gas cars and giving them a longer range.

Backup Documents: *President Obama Announces $2.4 Billion in Grants to Accelerate the Manufacturing and Deployment of the Next Generation of U.S. Batteries and Electric Vehicles, The Recovery Act: Transforming the American Economy through Innovation*

Box 6: A123 Systems – A Case Study

A smart innovation strategy recognizes that today's nascent invention in a university laboratory could be tomorrow's start-up company raising venture capital and eventually growing into a large company that employs hundreds of workers and addresses a key national priority.

Take the case of hybrid and electric vehicles. For these vehicles to achieve the market share necessary to significantly reduce carbon emissions, they must have batteries that are powerful, efficient, cost effective and safe. Where will these technologies come from and who will fund the entrepreneurs who develop them into commercial products and create jobs? Moreover, will this process of innovation and entrepreneurship happen in the United States or elsewhere?

Understanding the success of A123 Systems and the effective yet limited role of public policy in facilitating its development is particularly instructive. A123 Systems is a leading producer of the lithium-ion batteries that power hybrid and electric vehicles. The firm was founded in 2001 as a spin-off from MIT, enabled by a $100,000 SBIR grant from the Department of Energy. Improving the process by which promising university research is commercialized is a key component of our Innovation Strategy.

Buoyed by its early success transitioning from the laboratory to an entrepreneurial venture, the firm went on to raise more than $300 million in venture capital from venture investors like Sequoia Capital and corporations like GE and Motorola. It also received a $5 million loan from the Massachusetts Clean Energy Center.

A123 went public in September 2009, raising $380 million, and today it has approximately 1,700 employees. In December 2009, A123 obtained nearly $250 million from the Department of Energy to build a manufacturing plant in Michigan. The Administration's Innovation Strategy aims to catalyze entrepreneurial activity to address key national priorities, whether it nurtures firms like A123 in the clean energy sector or other innovative companies in biomedical sciences or nanotechnology.

Backup Documents: *A123 Systems–Revitalizing American Manufacturing, Remarks by the President in Phone Call to Recovery Act Advanced Battery Grant Recipient, A123 Systems in Livonia, MI.*

Promote Energy Efficient Industries

- Increased Demand to Improve Fuel Efficiency and Reduce Oil Dependence | EPA, NHTSA, and DOE

Beyond supporting investments in new vehicle technologies, the government is raising demand for innovation through new fuel efficiency standards and spurring private sector innovation in alternative fuels. In April 2010, EPA and NHTSA announced a joint Final Rule to establish greenhouse gas (GHG) emissions standards and CAFE standards respectively, for model year 2012-2016 light-duty vehicles. In September 2010, EPA and NHTSA issued a Notice of Intent to begin developing new standards for greenhouse gases and fuel economy for light-duty vehicles for the 2017-2025 model years. The two agencies are also focused on medium- and heavy-duty vehicles, and in October 2010, announced a proposed rule to, for the first time ever, establish GHG emissions standards and CAFE standards for such vehicles for model years 2014-2018. In parallel to these rule-makings, the Recovery Act provided the Department of Energy with the opportunity to invest nearly $718 million in the future of biofuels and biopower.

Backup documents: *EPA Regulations and Standards, May 19, 2009: President Obama Announces National Fuel Efficiency Policy, Biomass Program Recovery Act Factsheet, Biomass Program–Bioindustry Creates Green Jobs, Growing America's Energy Future: Renewable Bioenergy, Biomass Program Recovery Act Factsheet, Biomass Program–Bioindustry Creates Green Jobs, Growing America's Energy Future: Renewable Bioenergy*

- Government Demand as Innovation Driver | The White House, DOE, and GSA

The federal government, as the single largest energy consumer in the U.S. economy, can play a critical role catalyzing demand for innovative energy technologies. In October 2009, President Obama signed an Executive Order that calls on agencies to cut the federal government's fleet petroleum use by 30% by 2020, and in January 2010 President Obama announced that the federal government would reduce its GHG pollution by 28% by 2020.

As part of these efforts, the General Services Administration is replacing 5,603 of the least fuel efficient cars and trucks in the federal fleet with fuel efficient hybrids. The U.S. Navy is driving demand for new fuels by committing to convert its planes, ships, and vehicles to 50% alternative fuels by 2020. The Department of Defense (DOD) is also using its installations as a test bed and market driver for the next generation of energy technologies. For example, DOD is testing a smart microgrid that will allow the Twenty-nine Palms Marine Corps base to operate off the grid if needed and reduce its energy costs when tied to the grid. At Camp Roberts in California, the Army is validating a new solar technology that exploits advances in nanotechnology to provide significant cost savings. And at the Great Lakes Naval Training Center in Chicago, the military is testing an innovative building management system that continuously compares design specifications to actual building performance to allow immediate corrective actions. For these and other energy technologies that prove their worth, DOD can use its significant buying power to help create a commercial market, much as the military did with jet engines, integrated circuits, and the Internet.

Backup documents: *President Obama Sets Greenhouse Gas Emissions Reduction Target for Federal Operations, President Obama Expands Greenhouse Gas Reduction Target for Federal Operations, GSA Doubles the Federal Hybrid Fleet, DOE Takes the Lead in Updating to Hybrids, The Department of Energy's Fleet Alternative Fueled Vehicle Acquisition Report Compliance with EPACT 1992 and E.O. 13423 in Fiscal Year 2007*

Invest in Advanced Vehicle Technology

The President's FY 2012 Budget proposes to make the United States the world's leader in manufacturing and deploying next-generation vehicle technologies, expanding funding for vehicle technologies by almost 90% to nearly $590 million and enhancing existing tax incentives. Recovery Act and prior year investments are already making progress on advanced technology vehicles through research initiatives like an ARPA-E grant to develop a battery that will go 300 miles on a single charge. The FY 2012 Budget will significantly broaden R&D investments in technologies like batteries and electric drives – including an over 30% increase in support for vehicle technology R&D and a new Energy Innovation Hub devoted to improving batteries and energy storage for vehicles and beyond. In addition, the President is proposing to transform the existing $7,500 tax credit for electric vehicles into a rebate that will be available to all consumers immediately at the point of sale.

Promote Energy Efficient Buildings

- Greater Energy Efficiency in the Commercial Buildings | DOE and SBA

As a key part getting 80% of our electricity from clean energy resources, the President is calling for new efforts to improve energy efficiency in commercial buildings across the country. The President's Better Buildings Initiative will make non-residential buildings 20% more energy efficient over the next decade by catalyzing private sector investment through a series of incentives to upgrade offices, stores, schools and other municipal buildings, universities, hospitals, and other commercial buildings. The President's Budget will propose to make American businesses more energy efficient through tax incentives, increased financing opportunities, and a competitive grant program for states and municipalities.

Accelerate Biotechnology, Nanotechnology, and Advanced Manufacturing

Complete DNA Sequencing for Major Diseases and Drive Sequencing Innovations

A decade after the publication of the first draft human genome (a culmination of years of effort by thousands of researchers), the Administration is investing in the sequencing of over 1,800 complete genomes, a more than 50-fold increase over the 34 genomes that have been sequenced to date by non-Recovery Act NIH funding (*See Figure 4*). This accomplishment will be spread across various studies, projects, and disease groups, and will lend a new level of insight into disease that was previously unattainable.

The Administration, through NIH, is also leading The Cancer Genome Atlas (TCGA), the largest and most comprehensive analysis of the molecular basis of cancer ever undertaken. TCGA aims to identify and catalog all of the relevant genetic alterations in 20 common types of cancer. The genomic information generated by TCGA may lead to new targets for cancer treatments, improvements in how cancer is diagnosed, and advances in personalized medicine, such as prevention strategies based on an individual's genetic makeup and treatment plans based on a patient's unique disease.

The promise of being able to compare entire human genomes is boundless. Today, however, the financial cost of doing so is very high and often prohibitive. Therefore, further investments are directed toward the goal of slashing the cost of DNA sequencing. NIH has already made great progress in this area and aims to reduce the cost of sequencing a human genome to $15,000 in 2012, a 40% reduction from the current cost and almost 80% less than the cost in 2009.

Backup Documents: *The Recovery Act–Transforming the American Economy Through Innovation*

Feed the Future

Agricultural development hinges on access to affordable, appropriate, and sustainable technologies that can improve food production, harvesting, storage, and distribution, and advance the health and safety of all citizens. In developing countries with largely agrarian populations, a vibrant and sustainable agricultural system is the very basis for broad economic development and stability. By promoting the exploration of biotechnology and other strategies to improve agricultural productivity, the Feed the Future Initiative will ensure the food security in the U.S. and globally.

Backup documents: *Feed the Future Initiative*

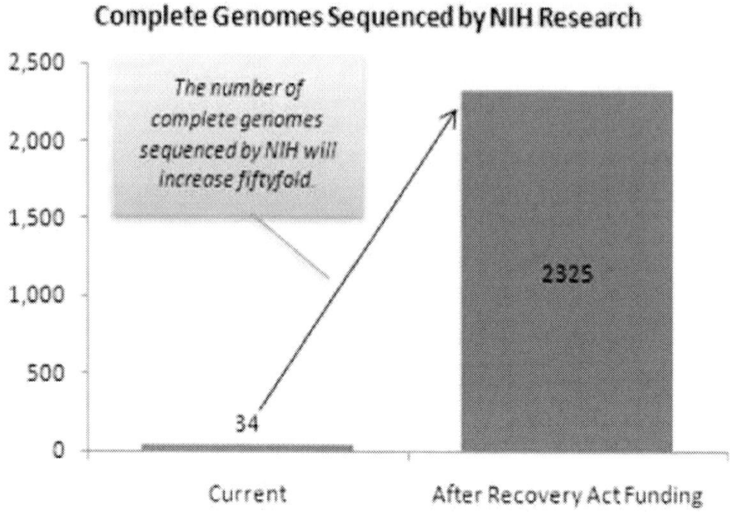

Figure 4.

Create Nanotechnology Solutions

Advances in nanotechnology can drive economic growth, create quality jobs, and address a broad range of national challenges. Examples of potential nanotechnology applications include smart anti-cancer therapeutics that target tumors without the devastating side effects of chemotherapy, solar cells as cheap as paint, and the next revolution in computing. Companies are already offering nanotechnologyenabled products with breakthrough capabilities in areas such as disease detection, lighter and stronger materials and next-generation batteries.

As indicated in the National Nanotechnology Initiative Supplement to the President's 2011 Budget, to accelerate nanotechnology development in support of the President's priorities and innovation strategy, NNI member agencies have identified areas ripe for significant advances through close and targeted program-level interagency collaboration:

- Nanoelectronics for 2020 and Beyond
- Sustainable Nanomanufacturing – Creating the Industries of the Future
- Nanotechnology for Solar Energy Collection and Conversion

Backup Documents: *Draft National Nanotechnology Initiative Strategic Plan 2010, No Input is Too Small: Comment on National Nanotechnology Initiative's Strategic Plan*

Launch Breakthrough Technologies for Advanced Manufacturing

A vibrant manufacturing base is critical for continued innovation, with important spillover benefits for the broader economy. While the manufacturing sector directly represents only 12% of U.S. GDP, manufacturing firms conduct 70% of private sector R&D. The manufacturing sector also helps train the next generation of skilled workers and supports the foundational knowledge necessary to produce process and product innovation.

Advanced manufacturing can be defined as manufacturing activities that bring substantial intellectual capital, in the form of new technologies, into the manufacturing process, resulting in a combination of lower cost, rapid customization, improved time-to-market, reduced waste, and increased quality. Advanced manufacturing technologies hold the potential to dramatically enhance the competitiveness of both existing and new manufacturing industries. The U.S. is already regarded as the leader in foundational knowledge in many of the technologies believed to have potential, including modeling and simulation, nanotechnology, information technology, and cyber-physical systems. However, to date, leadership in foundational knowledge has not translated into advanced manufacturing leadership.

The Administration is committed to support private sector investment in advanced manufacturing.

The Administration is launching private-public partnerships and investing in breakthrough research that can provide the foundation for future economic growth and competitiveness. The FY 2012 Budget increases investments at key science agencies, including the NSF, NIST laboratories, the Department of Energy Office of Science, and DARPA, to support U.S. leadership in developing advanced manufacturing technologies. The FY 2012 Budget also provides funding to initiate the Advanced Manufacturing Technology Consortia Program and sustain the Technology Innovation Program, public-private partnerships that will help spur innovation in new manufacturing products and processes.

Develop Breakthrough Space Capabilities and Applications

In June 2010, the President released his National Space Policy, which highlights the ever-increasing role that space systems, technologies, and activities play in our daily lives. Space technologies now help warn us of impending natural disasters, enable worldwide communications, facilitate global commerce, monitor global weather and climate, explore the solar system and universe, inspire our youth, provide global navigation and position location capabilities, and contribute to our security at home and abroad. The President is committed to continuing U.S. leadership in this arena by strengthening our space-related science and technology R&D initiatives, enhancing mutually beneficial international cooperation efforts, improving the state of the U.S. space industrial base, and taking steps to enable a robust, innovative, and competitive commercial space sector.

NASA will again be helping to lead the way in our space programs for the 21st century as it pursues a bold new approach to space science, exploration, and discovery. Under direction from President Obama and the U.S. Congress, NASA has been directed to establish a new program in advanced space technology R&D, to begin a new era of utilization of the International Space Station (ISS) by working with the private sector to expand American industry's role in human spaceflight, and to initiate the development of the next generation of space vehicles that will expand our human presence beyond Earth orbit and deeper into our solar system. These efforts will create thousands of new jobs as NASA expands its partnerships with industry, generate new markets for new products and discoveries derived from ISS utilization activities, expand our knowledge of our solar system and our place in it, and trigger a rush of new innovations and achievements.

In conjunction with NASA's new initiatives, the Obama Administration is continuing its efforts to advance U.S. capabilities in other space sectors, as highlighted in the President's National Space Policy. For example, the U.S. Department of Defense is in the process of launching a new generation of global positioning satellites and services and continuing the development of the satellite navigation architecture. These new capabilities will advance satellite navigation and timing applications in a range of fields, including areas such as urban planning, agriculture, communications, air travel, highway safety initiatives, and many more. The Administration will also continue and improve a broad array of programs of space-based observation, research, and analysis of the Earth's land, oceans, and atmosphere. New initiatives are also underway in areas such as satellite remote sensing and communications, each of which can help improve our security and support new industries and markets.

Backup documents: *National Space Policy*

Drive Breakthroughs in Health Care Technology

The inefficiencies in our health care system raise costs and reduce the quality of care. Information technology has the potential to revolutionize our health care systems, creating technological platforms to reduce costs, reduce errors, and increase the quality of care By breaking down barriers between health service providers, health IT can integrate our health markets to attract scalable, competitively tested private sector innovation.

Expand the Use of Health IT

- Health IT Adoption | Office of the National Coordinator (ONC) for Health Information Technology

Information technology has the potential to revolutionize our health care systems, creating technological platforms to increase efficiencies, reduce errors, and improve the quality of care. By breaking down barriers between health service providers, health IT can integrate our health markets to attract scalable, competitively tested private sector innovation. The Office of the National Coordinator (ONC) for Health Information Technology is working to realize this vision through numerous initiatives.

- The Medicare and Medicaid Electronic Health Record (EHR) Incentive Programs will provide incentive payments to eligible professionals and hospitals who qualify as "meaningful users" of certified EHR technology. According to surveys released in January of 2011 by the American Hospital Association and the National Center for Health Statistics, approximately 80% of eligible hospitals and 40% of office-based physicians intend to participate in the program.
- The Health Information Technology Extension Program establishes Regional Extension Centers to accelerate adoption of EHRs. Starting in 2010, $677 million has been awarded to create 62 Regional Extension Centers, and an additional $50 million will be invested in establishing the national Health IT Research Center. The Regional Extension Centers aim to support at least 100,000 primary care providers in achieving meaningful use of EHRs.
- The Nationwide Health Information Network (NwHIN) develops standards, services, and policies that enable secure health information

exchange over the Internet. The NwHIN will provide a foundation for the exchange of health IT across diverse entities, within communities, and across the country, providing a platform to attract and adopt system-wide innovation.

- The State Health Information Exchange Cooperative Agreement Program funds states' efforts to rapidly build capacity for exchanging health information across the health care system both within and across states. In March 2010, ONC announced a total of 56 awards – to states, eligible territories, and qualified State Designated Entities – amounting to over $547 million.
- The Beacon Community Program began providing $250 million in May 2010 to support 17 communities with expected EHR adoption rates significantly above national averages. The program will help accelerate and demonstrate the role of health IT in improving Americans' health and the performance of the health care providers who serve them.

Backup Documents: *HITECH Priority Grants Program–Factsheet, HITECH Priority Grants Program Homepage, State Health Information Exchange Cooperative Agreement Program–Factsheet, State Health Information Exchange Cooperative Agreement Program: Frequently Asked Questions, State Health Information Exchange Toolkit, National Institutes of Health–Mobile Health Homepage, Beacon Community Cooperative Agreement Program–Factsheet, Vice President Biden, HHS Secretary Sebelius Announce Selection of 15 Health IT Pilot Communities through Recovery Act Beacon Community Program,* Nationwide Health Information Network (NHIN) Homepage

- Breakthroughs in health IT technology | Office of the National Coordinator (ONC) for Health Information Technology

The Strategic Health IT Advanced Research Projects (SHARP) Program funds potentially game-changing advances to address well-documented problems that have impeded adoption of health IT. In March 2010, four $15 million dollar awards were announced to four academic institutions. Awardees will engage in research and product development to address barriers to health IT adoption in four key areas:

- Security and health information technology

- Patient-centered decision making support
- Health care application and network design
- Secondary use of EHR information

These academic partners will integrate input from providers, patients, academics, industry, and government stakeholders to translate their research into patient-centered products and services. This project will yield fundamental improvements within and between project areas to help make an electronic health record available for all Americans by 2014.

Backup Documents: *SHARP Program Factsheet,* SHARP Program Frequently Asked Questions Website

Enable Innovation in Medical Technologies and Medical Care

- Council on Medical Device Innovation and Transparency Initiative | U.S. FDA

The U.S. Food and Drug Administration is working to accelerate the development of medical device technologies. In June 2009, the FDA launched a "Transparency Initiative" which will improve the market's understanding of the approval process and encourage innovation. In January 2010, the FDA announced the creation of the Council on Medical Device Innovation, designed to encourage innovations that will address unmet public health needs. The multi-agency council will facilitate medical technology innovations by identifying the most important unmet public health needs, delineating the barriers to the development of technologies to treat these needs, and addressing obstacles to patient access.

Backup Documents: *FDA–Medical Device Development Process Presentation, FDA Public Workshop: Identifying Unmet Public Health Needs and Facilitating Innovation in Medical Device Development*

- Qualified Therapeutic Discovery Project Program | Internal Revenue Service and Health and Human Services

The Qualified Therapeutic Discovery Project Program provided tax credits and grants to small firms in 2009 and 2010 that show significant potential to produce new and cost-saving therapies, support U.S. jobs and increase U.S.

competitiveness. Applicants were required to have their research projects certified as eligible for the credit or grant. Health and Human Services, through the National Institutes of Health, reviewed the applications from a scientific perspective, and the Internal Revenue Service made the final determination of the awards.

Backup Documents: Program Homepage, *IRS guidance*

- Health Care Innovation Through Reduced Cost | Health and Human Services

Over time, the Affordable Care Act will reduce costs throughout the health care system. The bill creates competition among private-sector insurance companies in newly-created insurance exchanges. Most importantly, in the current fee-for-service system, manufacturers, hospitals, and other providers are generally reimbursed based on the volume and price of treatment, creating little motivation to innovate to contain costs. With the Affordable Care Act, delivery system reforms through accountable care organizations, bundling payments, and pressure to reduce hospital-acquired conditions and readmissions will all provide incentives for physicians, hospitals, and other providers to improve quality while lowering costs.

Projections by the Centers for Medicare and Medicaid Services find that the law will reduce the annualized growth rate of Medicare spending between 2012 and 2019 by 1.4 percentage points. An analysis by the Congressional Budget Office found that private-sector premiums for individuals will fall between 7 and 10 percent for a given amount of coverage.

Cost containment will provide strong incentives for innovation and the development of technologies (*see Box 7*) that improve the quality of care, saving lives and lowering costs by wringing inefficiency out of the health care system. In addition, reform will stimulate innovations in the processes of care and patient flow. A number of these innovations have occurred, but with health care reform the pace of innovation will accelerate.

Backup documents: *GAO Report on Hospital Flow, HHS Healthcare Related Infections Initiatives, Hospital Checklist Reduces Infections, GAO Report on Advanced Imaging*

Box 7: Innovations in Health Care

The innovations in health care that the Administration hopes to drive by containing costs go far beyond the important – but traditional – view of medical innovation as new drugs or approaches to treatment. Medical innovations range from redesigned devices to prevent errors that cost lives to changes in the workflow of hospitals and doctor's offices to the rapid accessibility of patient data. This box describes some of the technologies that most patients – and even many doctors – may not imagine are possible, yet would be scaled up within a system that rewards innovations by reducing costs:

- **Devices designed for efficiency and lower cost.** Curbing costs will require manufacturers to innovate to provide quality for less. For example, device designers have developed a resuscitation "crash cart" that splits into three pieces with dedicated equipment for each part of a medical team to provide care quickly and accurately in crisis situations. Designers have created lightweight, simpler plastic clamps to guide orthopedic surgeons in knee opera-tions, replacing cumbersome, overweight, metal instruments. The innovative guides improved accuracy because surgeons found them easier to employ, rather than operating freehand instead of using the older, bulkier devices. The current medical marketplace does not reward these types of innovative and efficient designs.
- **Safer medical equipment.** Safety can be embedded in the architecture of medical equipment itself. A 2006 survey found that 16% of hospitals had experienced a medical tubing error, such as attaching a tube delivering liquid food into a line leading to a vein – resulting in agony and death. Tubing designed so that it will not fit if it is connected improperly is an example of a medical equipment innovation that would reduce costs, prevent errors, and save lives.
- **Redesigned monitoring devices.** Patient monitoring devices—from those reporting vital signs to diagnostic tools —often have complex displays, and confusing ports for input and output. In high-pressure situations, these devices can hinder the effectiveness of nurses and doctors.

Better industrial design would yield a simpler interface with clear labels and display of patient information for quick reference. Designers have created diagnostic devices that improve safety as well, such as a blood pressure cuff with a Velcro-detachable, hygienic "wipe clean" surface, and a timestrip for intravenous lines that gradually fills with ink to provide a visual guide for how long a line has been inserted.

- **New office and hospital flow processes.** Management of the flow of patients between different arms of hospitals and within doctors' offices is an area ripe for innovation, and could be augmented with new technology. According to the Government Accountability Office (GAO), hospitals have improved the flow out of emergency departments and into inpatient beds by forming teams, which meet several times a day, to carefully monitor entering patients and opening beds. Hospitals have partnered with states and cities to establish diversion programs to reduce the burden on ER departments of patients who require treatment, but not emergency care. These processes save costs and improve quality, but are not rewarded under a system with high reimbursements for ER visits and payments based on volume.

• **Less expensive diagnostic tests.** According to the GAO, from 2000 through 2006, Medicare spending for imaging services paid for under the physician fee schedule more than doubled—increasing to about $14 billion. Spending on advanced imaging, such as CT scans, MRIs, and nuclear medicine, rose substantially faster than other imaging services such as ultrasound, X-ray, and other standard imaging. The Affordable Care Act takes steps to realign and rein in this growth in excess spending, driving providers towards effective and efficient diagnostic tests.

• **More health data.** As described elsewhere in the report, the Recovery Act makes large investments in health IT and electronic medical records. The Affordable Care Act drives new pricing and performance transparency of the health-insurance and health care-provider markets. The President's open government initiative continues to publish a growing array of health care data sets online and makes them freely available. Mining and sharing data from these sources will augment the implementation of all of the innovative technologies outlined above, by allowing the identification of high-risk individuals and good practices to address their needs.

• **A simple checklist for preventing infections.** Health care-associated infections (HAIs) exact a significant toll on human life. They are among the leading causes of preventable death in the United States, accounting for an esti-mated 1.7 million infections and 99,000 associated deaths in recent years. It is estimated that health care-associated infections cause $28 to $33 billion in excess health care costs each year. A simple, 5-point checklist developed at Johns Hopkins and implemented in hospitals in Michigan—which includes basic steps like washing hands, cleaning the skin, and wearing the proper mask, hat, gown and gloves—was found to reduce HAIs to nearly zero. In the current health care system, with payment based on volume rather than quality, these types of innovations in the structure of delivery are not rewarded.

Create a Quantum Leap in Educational Technologies

Advance Development of Educational Technologies

The Opportunity

There is growing evidence that technology can improve learning outcomes and reduce the time needed for students of all ages to gain new knowledge and skills by personalizing the learning experience. Furthermore, technology can increase the rate of improvement in education by enabling rapid, low-cost experimentation to better evaluate new methods and accelerate the diffusion of best practices. With relatively widespread access to broadband, cloud computing, digital devices and software, the technological conditions are increasingly ripe for the development of advanced educational technologies.

Additionally, learning technologies have the potential to be a rapidly growing industry internationally. As developing countries seek to educate broader segments of their population, they may turn to learning technologies. Many countries, such as India and China, have not invested in the brick-and-mortar infrastructure that we have in the United States, and they may seek less expensive, more flexible approaches to learning. With appropriate leadership, the United States has an opportunity to lead in this new and expanding market, but we must move rapidly to secure our leadership or other countries will close the gap.

Government's Role

The government's role in stimulating and facilitating the development of this sector could take several forms. In the education sector, where there is potential for widespread adoption of these technologies, demand is currently fragmented and insufficient to encourage the private sector to make substantial investments. As it has done successfully in other sectors of national importance (for example, semiconductors or biotechnology), the government is currently exploring limited and strategic investments in educational R&D that will catalyze the emerging education technology industry and secure American leadership in a global sector – upfront investments that many educational institutions and companies would not be able to make on their own but that will produce benefits that accrue to the entire nation.

The President has already demonstrated his commitment to educational technology and the Department of Education recently released a National Educational Technology Plan, which is a five-year action plan for using technology to improve student learning, accelerate and scale up the adoption of effective practices, and use data and information for continuous improvement. Across the government there are initiatives to support improved learning outcomes through the use of technology, from the National Science Foundation launch of "Cyberlearning Transforming Education" to a variety of DARPA, DOE, and Navy programs in this area. Given the opportunities in this sector the government will seek to promote the development and adoption of educational technology in the following ways:

- Support federal research, development, demonstration and evaluation that will improve our understanding of how and in what contexts educational technology can improve student achievement.
- Provide immediate opportunities for the creation of new learning technologies, including through support for the development of specific applications and courseware through interagency procurement agreements.
- Develop "pull" mechanisms to accelerate the market for advanced learning technologies, working with the Department of Defense Education Activity (DoDEA) schools as platforms for innovation and as early "customers" for new effective approaches.
- Stimulate private sector investment in the sector—call on companies and philanthropic organizations to supplement federal government spending on K–12 STEM education, building on the successful Investing in Innovation (i3) Fund.

There are a number of opportunities to catalyze market opportunities and make transformative investments, with an Advanced Research Projects Agency for Education (ARPA-ED) as a flagship leader in this field. Modeled after the Defense Advanced Research Projects Agency (DARPA), the government department that has supported the creation of the Internet, GPS, stealth technology, and other revolutionary innovations, the creation of ARPA-ED will drive innovation in the area of educational technology. With funding in Fiscal Year 2012, ARPA-ED will:

- Sponsor the synthesis and vetting of the nation's public and private research and development efforts with the explicit purpose of identifying breakthrough development opportunities and shaping the next wave of R&D.
- Invest in the development of new education technologies and digital learning materials.

- Identify and transition the best and most relevant R&D with possible applications in education from other federal agencies.

The Goal

ARPA-ED would fund bold programs and increase entrepreneurship and innovation in the field of education. The end goal is to develop transformative, game-changing education technologies – technologies that will be interoperable and build strategically upon one another to achieve progress at scale.

Backup documents: *National Educational Technology Plan*

End Notes

[1] See Angus Maddison, The World Economy: Historical Statistics, OECD Publishing, (2006)

[2] See Robert W Fogel, "Nutrition and the Decline in Mortality since 1700: Some Preliminary Findings" in Long-term Factors in American Economic Growth, Engerman and Gallman eds (1986), and the review by Michael Haines, "The Urban Mortality Transition in the United States, 1800-1940", National Bureau of Economic Research Historical Working Paper No 134 (2001)

[3] Robert D Atkinson & Scott M Andes, The Atlantic Century: Benchmarking EU and U S Innovation and Competitiveness, The Information Technology & Innovation Foundation (2009), available at http://www itif org/ publications/atlantic-century-benchmarking-eu-and-us-innovation-and-competitiveness

[4] See, e g , Douglas North and Robert Thomas, The Rise of the Western World: An Economic History, Cambridge: Cambridge University Press (1973); and Charles Jones, Was an Industrial Revolution Inevitable? Economic Growth over the Very Long Run, 2(1) Advances in Macroeconomics (2001)

[5] See also William D Nordhaus, "Schumpeterian Profits in the American Economy: Theory and Measurement", Yale University , 2004

[6] Federal Communications Commission, "Connecting America: The National Broadband Plan (NBP)," xi

[7] Ibid

Strategies for American Innovation
Editor: Zachary A. Monroy

ISBN: 978-1-61324-247-6

Chapter 2

A STRATEGY FOR AMERICAN INNOVATION: DRIVING TOWARDS SUSTAINABLE GROWTH AND QUALITY JOBS

Executive Office of the President
National Economic Council
Office of Science and Technology Policy

History should be our guide. The United States led the world's economies in the 20th century because we led the world in innovation. Today, the competition is keener; the challenge is tougher; and that is why innovation is more important than ever. It is the key to good, new jobs for the 21st century. That's how we will ensure a high quality of life for this generation and future generations. With these investments, we're planting the seeds of progress for our country, and good-paying, private-sector jobs for the American people."

-President Barack Obama, August 5, 2009

EXECUTIVE SUMMARY

Since taking office, President Obama has taken historic steps to lay the foundation for the innovation economy of the future. The Obama Innovation Strategy builds on well over $100 billion of Recovery Act funds that support innovation, additional support for education, infrastructure and others in the Recovery Act and the President's Budget, and novel regulatory and executive order initiatives. It seeks to harness the inherent ingenuity of the American ... ensure that the next expansion is more solid, broad-based, and beneficial than previous ones. It focuses on critical areas where sensible, balanced government policies can lay the foundation for innovation that leads to quality jobs and shared prosperity. It has three parts:

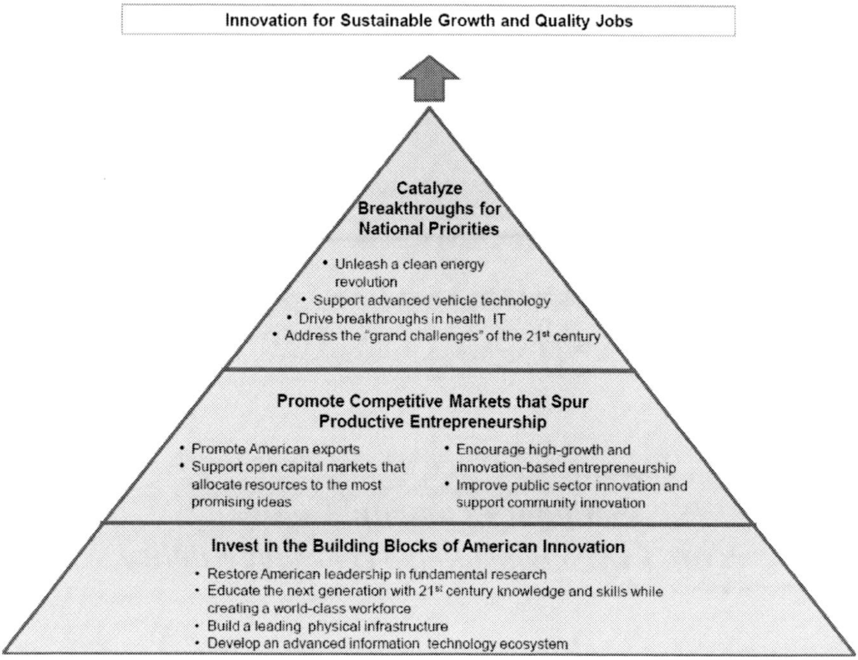

1. Invest in the Building Blocks of American Innovation. We must first ensure that our economy is given all the necessary tools for successful innovation, from investments in research and development to the human, physical, and technological capital needed to perform that research and transfer those innovations.

- **Restore American leadership in fundamental research.** President Obama implemented the largest increase in basic R&D in history, which will lay the foundation for new discoveries and new technologies that will improve our lives and create the industries of the future.
- **Educate the next generation with 21st century knowledge and skills while creating a world-class workforce.** President Obama has proposed initiatives to dramatically improve teaching and learning in K-12 education, expand access to higher education and training, and promote student achievement and careers in STEM (Science, Technology, Engineering, and Mathematics) fields.

- **Build a leading physical infrastructure.** Through the Recovery Act, the President has committed to a historic investment in our nation's roads, bridges, transit, and air transportation networks to connect our people and our businesses.
- **Develop an advanced information technology ecosystem.** For America to lead the world in the technologies of the future, President Obama believes that all Americans must have affordable 21st century access to the Internet.

2. Promote Competitive Markets that Spur Productive Entrepreneurship. It is imperative to create a national environment ripe for entrepreneurship and risk taking that allows U.S. companies to be internationally competitive in a global exchange of ideas and innovation. Through competitive markets, innovations diffuse and scale appropriately across industries and globally.

- **Promote American exports.** Exports will play an increasingly critical role in the future of the American economy, and the President's plans will ensure fair and open markets for American producers.
- **Support open capital markets that allocate resources to the most promising ideas.** Open capital markets are one of the greatest strengths of the American economy, and the President is committed to making sure these markets work.
- **Encourage high-growth and innovation-based entrepreneurship.** The Obama Administration believes it is essential that entrepreneurs continue to create new and vibrant businesses that lead to new jobs and economic growth.
- **Improve public sector innovation and support community innovation.** Innovation must occur within all levels of society, including the government itself. The Obama Administration supports the broad adoption of community innovations that work and is committed to making government perform better and more efficiently, including by working more openly.

3. Catalyze Breakthroughs for National Priorities. There are certain sectors of exceptional national importance where the market is unlikely to produce the desirable outcomes on its own. These include developing alternative energy sources, reducing costs and improving lives with health IT, and manufacturing advanced vehicles. In these industries where markets may fail on their own, government can be part of the solution.

- **Unleash a clean energy revolution.** Historic investments in smart grid, energy efficiency, and renewable technologies like wind, solar, and biofuels will help unleash a wave of ingenuity and progress that creates jobs, grows our economy, and ends our dependence on oil.
- **Support advanced vehicle technologies.** Record battery grants announced last month are part of a concerted effort to place the U.S. on the cutting edge of advanced vehicle technology, from electric cars to biofuels to advanced combustion.
- **Drive innovations in health care technology.** The President's health IT initiative is designed to drive technological innovation that will help prevent medical errors, improve health care quality, reduce costs, and cement U.S. leadership of this emerging industry.
- **Harness science and technology to address the "grand challenges" of the 21st century.** The President's commitment to science and technology will allow the United States to set and meet ambitious goals, such as educational software that is as effective as a personal tutor and smart anti-cancer therapeutics that deliver drugs only to tumor cells.

I –Problems with the Bubble-Driven Growth of the Past

A Strong Economy, but too Reliant on Precarious, Bubble-Driven Growth, Is Unsustainable

Despite American economy's historic strength, our economic growth has rested for too long on an unstable foundation. Explosive growth in one sector of the economy has provided a short-term boost while masking long-term weaknesses. In the 1990s, the technology sector climbed to new heights. The tech-heavy NASDAQ composite index rose over 650 percent between 1995 and 2000, but then lost two-thirds of its value in a single year, triggering a painful recession.

After the tech bubble burst, a new one emerged in the housing and financial sectors. During the course of the decade, the formula for buying a house changed: instead of saving to buy their dream house, many Americans found they could take out loans that by traditional standards their incomes could not support. The financial sector willingly propped up real estate prices, funneling money into real estate and finding innovative ways to spread the credit risk throughout the economy. From 2000 through 2006, house prices doubled while the financial sector grew to account for fully 40 percent of all corporate profits.

This too proved to be unsustainable. House prices lost a quarter of their value in two and a half years. The housing decline and accompanying stock market collapse wiped out over $13 trillion in wealth in 18 months. The bursting of the bubble based on inflated home prices, maxed-out credit cards, over-leveraged banks, and overvalued assets wreaked havoc on the real economy, triggering what is expected to be the longest and deepest recession since World War II and driving the unemployment rate to its highest level in a quarter century.

This type of growth isn't just problematic when the bubble bursts, it is not entirely healthy even while it lasts. Between 2000 and 2007 the typical working-age household saw their income decline by nearly $2,000. As middle-class incomes sank, the incomes of the top one percent skyrocketed. This phenomenon has a number of causes, but among them were the rising asset prices and the proliferation of financial sector profits.

A Short-Term Focus Has Neglected Essential Fundamental Investments

A short-term view of the economy masks underinvestments in essential drivers of sustainable, broadly-shared growth. It promotes temporary fixes over lasting solutions. This is patently clear when looking at how education, infrastructure, healthcare, energy, and research – all pillars of lasting prosperity – were ignored during the last bubble.

Too many children are not getting the world class education they deserve and need to thrive in this new innovative economy.

- Despite research documenting that quality matters greatly in early childhood education settings and that investments in high-quality early learning have the highest potential rates of return, the Federal government lacks the level of investment needed to transform the quality of, and enhance access to, early education for our youngest children. Studies show a school readiness gap as early as kindergarten –and as wide as 60 points –between children from the highest socio-economic background and their less affluent peers.
- We have neglected to provide our children with the rigorous curriculum and instruction needed to prepare them for college and career. By the end of high school, African American and Latino students have math and reading skills equivalent to those of 8th grade white students. Across the nation, the students with the greatest need for a qualified and effective teacher are also exactly those students

most likely to be taught by teachers who lack sufficient background in the subject they teach.

- The problems persist when students look toward continuing their education past high school. The average tuition and fees at public, four-year institutions rose 26 percent between the 2000-2001 school year and the 2008-2009 school year. As a result, while 94 percent of high school students in the top quintile of socioeconomic status continue on to post-secondary education, barely half of those in the bottom quintile do so.
- Given rising costs of four-year institutions, many Americans are turning to community colleges for quality higher education. Yet the Federal government has historically under-invested in community colleges, giving them one-third the level of support per full-time equivalent student that it gives to public four-year colleges.

Our physical and technological infrastructure has been neglected, threatening the ability of American businesses to compete with the rest of the world.

- The American Society of Civil Engineers grades our country's physical infrastructure as a "D." In 2007, drivers on our clogged highways and streets experienced over 4.2 billion hours of delay and wasted 2.8 billion gallons of fuel.
- The United States once led the world in broadband deployment, but now that leadership is in question. Wireless networks in many countries abroad are faster and more advanced than our own.
- Our electrical grid is still constructed around the same model employed immediately after World War II. Power interruptions and outages cost American individuals and businesses at least $80 billion each year.

Health care costs have been allowed to spiral out of control, squeezing individuals and businesses at a time when they are feeling pressure on all sides.

- Since 2000, health insurance premiums have increased about 60 percent, 20 times faster than the average worker's wage.
- At the same time, the number of uninsured Americans has jumped by 7 million to 46 million.
- Overall, healthcare is consuming an increasing amount of our Nation's resources. In 1970, healthcare expenditures were 7 percent of

GDP; now they are 16 percent of GDP; at this rate they will hit 20 percent of GDP by 2017.

Our economy has remained dependent on fossil fuels, exposing consumers and businesses to harmful price shocks, threatening our economic and national security, and resulting in a missed opportunity to lead the clean energy economy of the future.

- Between 1999 and 2004, the production tax credit for renewable energy was allowed to expire on three separate occasions. In each subsequent year (2000, 2003, and 2004) new wind capacity additions in the U.S. fell by more than 75 percent from the year before.
- Instead of focusing on finding ever more fossil fuels, other countries made aggressive investments in renewable energy, creating jobs and growing domestic energy sources.

Furthermore, we have compounded the problem by ignoring essential investments in high-technology research that will drive future growth.

- Over the last four decades, Federal funding for the physical, mathematical, and engineering sciences has declined by half as a percent of GDP (from 0.25 percent to 0.13 percent) while other countries have substantially increased their research budgets.

Despite this underinvestment in key drivers of growth, the American economy remains the most dynamic, innovative, and resilient in the world. We still have world-class research universities, flexible labor markets, deep capital markets, and an energetic entrepreneurial culture. Americans are twice as likely as adults in Europe and Japan to start a business with the intention of growing it rapidly. We must redouble our efforts to give our world-leading innovators every chance to succeed. We cannot rest on our laurels while other countries catch up.

II – A Vision for Innovation, Growth, and Jobs

Innovation Is Necessary to Fuel Our Recovery

Amidst the worst recession since the great depression, the Administration's initial economic objective has been to rescue our economy. We have taken – and will continue to take – bold and aggressive steps to stabilize the financial system, jumpstart job growth, and get credit flowing again.

But as the economy begins to stabilize, we must move on from rescue to recovery. Reflecting on the lessons of the past, we must rebuild a new foundation for durable, sustainable economic growth.

Innovation and investment must be a pillar of this new foundation. The basis of competition and the nature of the economy have changed, and we must change with them. Twenty years ago, the United States was losing domestic manufacturing firms and was competing with other countries to sell its goods. Now, manufacturing and services have merged, knowledge is a key factor of production, and services we thought could only be provided in particular countries are available anywhere. We need new ideas to provide Americans with new jobs, new services that take advantage of our globally interconnected world, and new skills that improve our manufacturing capabilities (See Box 1). Other countries understand that innovation is fundamental to their economic well-being and are finding new ways to advance their innovation agendas. We can be even more ambitious, even more successful, and even more focused on building the essential sidewalks of innovation.

Box 1. The Importance of a National Innovation Strategy

Fundamentally, innovation is the development of new products, services, and processes. In our increasingly interconnected and globally competitive world economy, unleashing innovation is an essential component of a comprehensive economic strategy. As global competition erodes the return to traditional practices, the key to developing more jobs and more prosperity will be to create and deploy new products and processes. Put another way, the greatest job and value creators of the future will be activities, jobs, and even industries that don't exist yet today. The countries that catalyze their development will reap the greatest rewards.

Innovation is essential for creating new jobs in both high-tech and traditional sectors. In recent years, innovation has led to new jobs in high-tech and advanced manufacturing sectors as diverse as aerospace, nanotechnology, life sciences, and alternative energy. At the same time, innovations ripple through the economy, creating jobs for workers installing broadband networks, manufacturing biopharmaceuticals, and building advanced infrastructure.

A more innovative economy is a more productive and faster growing economy, with higher returns to workers and increases in living standards. America's average standard of living will double every 23 years if innovation catalyzes annual productivity growth of three percent, but it will take 70 years if productivity growth is only one percent. Currently, the U.S.

enjoys a significant productivity advantage: one study calculates that the average productivity advantage of the United States over all other OECD countries as a group accounted for a full three quarters of the per capita income advantage the U.S. enjoys. Continued innovation in products, business practices, and technology is essential for extending our productivity gains.

Innovation is also crucial for maintaining the dynamism and resilience of our economy. Future challenges are impossible to predict, but what is certain is that an economy better able to switch gears, innovate solutions, and re-deploy old activities, jobs, and industries will be least susceptible to adversity.

Finally, innovation is itself the key to meeting some of the greatest challenges facing our nation and the world. It will be pivotal to ending our dependence on fossil fuels, helping Americans live longer, healthier lives, and protecting our freedom and our troops both at home and abroad.

Innovation is the key to global competitiveness, new and better jobs, a resilient economy, and the attainment of essential national goals. A strategy is clearly needed to direct our government's funding and regulatory decisions in order to capture the innovation opportunity.

Our Vision for American Innovation

Our vision of America's future is one where prosperity is built by skilled, productive workers and sound investments that will spread opportunity at home and allow this nation to lead the world in the technologies, innovation and discoveries that will shape the 21st century.

Innovation will create new jobs and catalyze broadly shared economic growth. The lives of every American will improve as innovations diffuse and scale throughout the economy, leading to breakthroughs in health, education, energy, transportation, information, and much more. We can set and meet grand challenges such as developing solar cells as cheap as paint, building anticancer drugs that spare healthy cells, and fitting the contents of the library of congress on a device the size of a sugar cube.

Sustained innovation will drive a dynamic evolution in the nation's workforce towards better paying jobs. American workers will continue to lead across a broad range of industries and sectors old and new, and will prosper accordingly. Workers making innovation-induced shifts to new jobs within and across industries will receive the transition and training support necessary to ensure no one is left behind.

The American economy is and will remain highly diversified. To a large ... these jobs of the future will be spread across major industries in a similar distribution to today's economy. As the Council of Economic Advisors described in a July report, one forecast that uses the most recent data available to project employment growth finds that the distribution of jobs across industries in the economy of 2020 will strongly resemble the distribution of 2008. Health and education services will see the most significant growth, while there will be proportionally fewer jobs in business and financial services as well as retail trade, in part because the growth in consumer spending is expected to slow. But by and large the picture is similar to today.

Box 2. The Transformation of the Semi-Conductor Industry

In the 1980s, the U.S. semiconductor industry lost its market share to Japanese competitors. But then it innovated its way back, replacing the old jobs in the dynamic-random-access-memory (DRAM) business with jobs producing microprocessors, digital signal processors, microcontrollers, and automotive semiconductors. Companies like Intel, Texas Instruments, and Motorola invested and succeeded, creating better jobs for hundreds of thousands of Americans. Throughout this shift to higher-value-added jobs, the total number of U.S. jobs in the industry held constant.

Of course, such forecasts are necessarily imperfect. They cannot capture the growth in industries that may not yet exist. For example, in the late 1980s there were no models that predicted the rapid growth of Internet-based information and computing services that now employ several million workers. Innovation in this sector –initially catalyzed and continuously supported by government investment –has made key contributions to our economy. Some experts estimate that the Internet adds as much as $2 trillion to annual GDP, over $6,500 per person.

Another shortcoming of these models is that they fail to capture many of the most important innovations and developments occurring in sub-industries. Yet the U.S. has a history of innovating towards higher-wage jobs within industries. The experience of industries as diverse as semi-conductors and boatbuilding illustrate this point amply (see Boxes 2 and 3).

We see an American future where this process of innovation in next-generation technologies and business ideas becomes pervasive, diffusing throughout the economy and generating better jobs and improved lives for all Americans. It is imperative that we turn this vision into a reality.

Box 3. Boat Builders along Maine's Eastern Coast

The Maine boat building industry is an example where innovation-led transformations occur on a smaller, more local scale. Maine boat builders have a 400-year heritage of skilled craftsmanship, but technological change was threatening to leave them behind. Instead, the boat builders have enthusiastically embraced cutting edge innovation in advanced composite technologies, replacing old jobs with better jobs in the same industry. Today, the Maine boat building industry produces a highly regarded product line that includes racing yachts, pleasure craft, workboats, and military vessels. As a result, wages in Maine's boat building industry have risen 19 in percent in real terms over the last decade while employment has risen 12 percent.

III –The Appropriate Role for Government

Framework for Government Involvement

While it is clear that a new foundation for innovation and growth is needed, the appropriate framework for government involvement is still debated. Some claim that the laissez-faire policies of the last decade approach the right strategy, and that the recent crisis was the result of too much rather than too little government support. This view calls for cutting government regulation and gutting public programs, hoping the market will take care of the rest.

However, the recent crisis illustrates that the free market itself does not promote the long-term benefit of society, and that certain fundamental investments and regulations are necessary to promote the social good. This is particularly true in the case of investments for research and development, where knowledge spillovers and other externalities ensure that the private sector will under-invest –especially in the most basic of research.

Another view is that the government must dominate certain sectors, protecting and insulating those areas thought to be drivers of future growth. This view calls for massive, sustained government investment supported by stringent oversight, dictating the type and direction of both public and private investments through mandates and bans.

But historical experience in this country and others clearly indicates that governments who try to pick winners and drive growth too often end up wasting resources and stifling rather than promoting innovation. This is in part due to the limited ability of the government to predict the future, but also more likely to favor backward looking industries than forward looking ones. In the United States such failures at picking winners and losers includes most prominently the Synthetic Fuel Corporation, a $20 billion Federal project in the 1980s that failed to provide the promised alternative to oil.

Therefore, we reject both sides of this unproductive and anachronistic debate. The true choice in innovation is not between government and no government, but about the right type of government involvement in support of innovation. A modern, practical approach recognizes both the need for fundamental support and the hazards of overzealous government intervention. The government should make sure individuals and businesses have the tools and support to take risks and innovate, but should not dictate what risks they take.

We propose to strike a balance by investing in the building blocks that only the government can provide, setting an open and competitive environment for businesses and individuals to experiment and grow, and by providing extra catalysts to jumpstart innovation in sectors of national importance. In this way, we will harness the inherent ingenuity of the American people and a dynamic private sector to generate innovations that help ensure the next expansion is more solid, broad-based, and beneficial than previous ones.

Examples of Successful Innovation Programs

We have been successful in the past, and we can be successful in the future. Take the example of DARPA, the historically innovative central research and development organization of the Defense Department. DARPA is tasked with maintaining U.S. technological superiority, and has a history of creating new industries in information technology and advanced manufacturing (see Box 4).

Box 4. DARPA's Successful Innovations

Thirty years ago, DARPA supported the creation of the internet's predecessor, ARPANET, despite discouragement from the private sector. Today, over 1.6 billion people use the Internet. DARPA's innovative research has created entirely new capabilities for the U.S. military such as stealth aircraft, GPS, the M-16 assault rifle, and night vision goggles. It has provided the foundation for new industries like optical networking, supercomputers, and design tools for computer chips. DARPA's commitment to high-risk, high-return research will help ensure America is prepared to meet the 21st technological innovations that will create new industries and improve people's lives.

Government support has also helped push America to the cutting edge in emerging fields such as nanotechnology, which involves engineering materials and devices on the atomic and molecular level (see Box 5).

The Administration is committed to strengthening and focusing investments in our world-class nanotechnology research and development pipeline; targeting support for nanotechnology transfer and facilitating commercial start-ups; and cross-disciplinary training and education of scientists and engineers in the new-generation workforce. This will enable us to capitalize on our investments and stay at the cutting edge of this rapidly growing technology.

This pattern of government support driving innovations that improve lives and catalyze industries is pervasive. Of the 88 U.S. entities that received "R&D 100 Awards" as the nation's best innovations in 2006, 77 had received government support. We must build on this record of successful support to build a new foundation for innovation and growth.

Box 5. The Growing Fields of Nanotechnology and Personalized Medicine

A nanometer is a billionth of a meter – or one hundred thousand times smaller than the diameter of a human hair. Nanotechnology promises to transform multiple industries: capturing and storing clean energy, developing next-generation computer chips, early detection of diseases, smart anti-cancer therapeutics that deliver drugs only to tumor cells, and enabling all-new approaches to a wide range of manufacturing activities, among many other examples.

While the commercial impact of nanotechnology to date has been limited primarily to nanomaterials applied to a range of consumer goods from healthcare and food products to textiles, automotive composites and industrial coatings, nanotechnology innovation is beginning to accelerate. The ten-year history of U.S. leadership in fundamental nanotechnology research and development under the National Nanotechnology Initiative has laid the crucial groundwork for developing commercial applications and scaling up production, creating demand for many new nanotechnology and manufacturing jobs in the near-term.

Nanotechnology is being applied in the developing medical engineering and personalized medicine industries. The practice of tailoring medical treatment to an individual's unique genetic make-up makes not only treatment, but early detection and prevention, more effective. It also reduces medical costs in cases where expensive treatments are unnecessary or futile.

Researchers are currently experimenting with nanotechnology to develop drugs capable of targeting a disease without triggering the body's natural immune response.

The federal government's support has been essential in the development of this technology and this industry. The NIH, a major supporter of medical research, saw its budget increase 163% from 1993 to 2003 before stagnating until this year. President Obama has reversed the recent trend with $10 billion in additional NIH funding in the Recovery Act and a pledge for more sustained increases going forward. This funding will help these emerging industries flourish.

IV –A STRATEGY FOR AMERICAN INNOVATION

For our communities and for our country to thrive in this new century, we need to harness the spirit of innovation and discovery that has always moved America forward. We must foster innovation that will lead to the technologies of the future – which will in turn lead to the industries and jobs of the future.

President Obama has already taken historic steps to lay the foundation for the innovation economy of the future. In the Recovery Act alone the President committed over $100 billion to support groundbreaking innovation with investments in energy, basic research, education and training, advanced vehicle technology, innovative programs, health IT and health research, high speed rail, smart grid, and information technology (see Figure 1). His commitment also includes broader support in the Recovery Act and in his FY2010 budget on initiatives from education to infrastructure. The President's commitment is not just limited to more government funding, but extends to important regulatory and executive order initiatives such as patent reform, coordinated fuel efficiency standards, net neutrality, permit policy for offshore wind farms, and naming the first ever Chief Technology Officer of the U.S. Government.

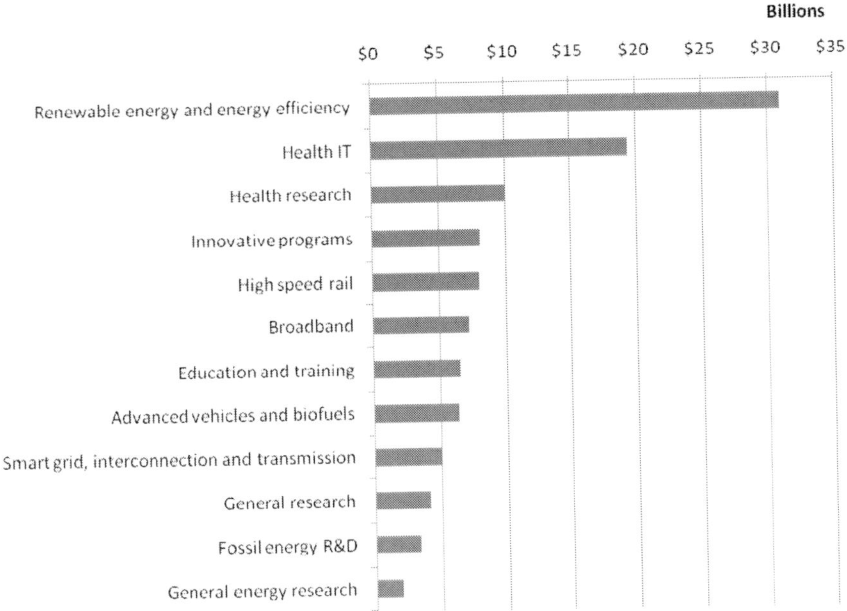

Figure 1. Innovation Funding in the Recovery Act.

These investments and initiatives are part of The Obama Innovation Strategy, which focuses on critical areas where sensible, balanced government policies can lay the foundation for innovation that leads to quality jobs and shared prosperity. It has three parts:

1) **Invest in the Building Blocks of American Innovation.** We must first ensure that our economy is given all the necessary tools for successful innovation, from investments in research and development to the human, physical, and technological capital needed to perform that research and transfer those innovations.
2) **Promote Competitive Markets that Spur Productive Entrepreneurship.** It is imperative to create a national environment ripe for entrepreneurship and risk taking that allows U.S. companies to be internationally competitive in a global exchange of ideas and innovation. Through competitive markets, innovations diffuse and scale appropriately across industries and globally.
3) **Catalyze Breakthroughs for National Priorities.** There are certain sectors of exceptional national importance where the market is unlikely to produce the desirable outcomes on its own. These include developing alternative energy sources, reducing costs and improving lives with health IT, and manufacturing advanced vehicles. In these industries where markets may fail on their own, government can be

1. Invest in the Building Blocks of American Innovation

A. Restore American Leadership in Fundamental Research

President Obama recognizes the fundamental role of government in fostering groundbreaking scientific and technological breakthroughs, and has committed resources and energy to ensure America leads the world in the innovations of the future.

- **Enact the Largest R&D increase in our nation's history.** With $18.3 billion in research funding, the Recovery Act is part of the largest annual increase in research and development in America's history.
- **Double the R&D budget of key science agencies.** The President's Budget proposed to double the research budgets of three key science agencies: the National Science Foundation, the Department of Energy's Office of Science, and the National Institutes of Standards and Technology. These investments will expand the frontiers of human knowledge and create the foundation for the industries and jobs of the future, such as the convergence of bio, info, and nanotechnologies. The Obama Administration will increase the impact of this investment by providing more support for high-risk, high-return research, for multidisciplinary research, and for scientists and engineers at the beginning of their careers.
- **Invest three percent of GDP in R&D**. The President has proposed a goal that as a country, we invest more than three percent of our GDP in public and private research and development. This will exceed the level achieved at the height of the space race.
- **Make the R&E tax credit permanent.** The President's Budget includes the full $75 billion cost of making the research and experimentation tax credit permanent. This will provide businesses with the greater confidence they need to invest, innovate, and grow.

B. Educate the Next Generation with 21st Century Knowledge and Skills while Creating a World-Class Workforce

America's high levels of education were an important contributor to rising living standards in recent years. For America to continue to lead the world in science and technology innovation, it must have the most knowledgeable and skilled workers in the world.

- **Reform public schools to deliver a complete and competitive education.** The President is committed to an education system that will prepare every child for success in a new, global economy. The Obama Administration is supporting the improvement of teaching and learning to ensure that students master world-class knowledge and critical skills for the 21st century; enhance and reward success in the teaching profession; drive innovation in America's classrooms; and expand successful models to improve outcomes for middle- and high-school students. The President's Race to the Top in America's schools will deliver a historic investment and challenge the states to design and enforce higher and clearer standards, attract and keep outstanding teachers in the classroom, and use effective approaches to turn America's lowest-performing schools around.
- **Restore America to first in the world in college graduates**. Colleges, universities, and their students are bedrocks of American innovation. President Obama has called for restoring America to first in the world in the proportion of college graduates by 2020. Between the Recovery Act and the 2010 budget, the Administration is seeking nearly $200 billion over the next decade for scholarships and tax credits to help students complete college. The Obama Administration is investing in innovative strategies to support college persistence, simplify the student aid application and reform the student loan program to shift billions in wasteful spending toward greater help for students to reach and complete college. The Administration has been actively working with Congress to fund these priorities by eliminating waste in the student aid program.
- **Improve America's Science, Technology, Engineering and Math (STEM) education.** STEM education is particularly important to America's future scientific progress and economic growth. As part of his effort to promote innovation in K-12 education, the President has challenged governors, philanthropists, scientists, engineers, educators and the private sector to join with him to dramatically improve achievement in STEM subjects. The President's Race to the Top fund aims to reinvigorate the teaching of STEM in America's classrooms and support advanced learning in these subjects, especially for women, girls and other underrepresented groups. The President is also committed to using the $4 billion Race to the Top fund to encourage states to put STEM at the center of their reform efforts. Finally, the Recovery Act provided a down payment toward the goal of tripling the number of NSF Graduate Research Fellowships in science and engineering.

- **Develop new career pathways in community college programs**. Given the innovation and advances coming in future years, jobs requiring at least an associate's degree are projected to grow twice as fast as those requiring no college experience. President Obama has proposed the American Graduation Initiative to produce 5 million more community college graduates by 2020, a key part of his college completion goal. The initiative would invest in promising reforms to raise graduation rates, tie courses to business needs, improve remedial education and strengthen transitions with high schools and four-year colleges. It would also leverage $10 billion in facilities modernization and repairs amongst other projects.
- **Design world-class online courses for post-secondary students**. The President is proposing to invest up to $500 million over the next 10 years to create world-class online courses available at community colleges for students to gain knowledge, skills and credentials. These courses will be freely available, and enable students to extend learning opportunities and successfully complete their course work. Colleges, universities, publishers, and other groups will be invited to compete to create state-of-the-art online courses that combine high-quality subject matter expertise with the latest advances in cognitive and computer sciences.
- **Improve the processing of high-tech visas.** In order to maintain our role as a global leader and convener of scientific conferences and other gatherings, the Obama Administration has worked to ensure foreign scientists and technological leaders can visit the United States to participate in important events while continuing to protect sensitive technologies.

C. Build a Leading Physical Infrastructure

The President has committed to a historic investment in our nation's roads, bridges, transit, and air networks to connect our people and our businesses.

- **Invest in our nation's roads, bridges, and mass transit.** The Recovery act provides $36 billion for infrastructure projects to improve our nation's highways and mass transit systems. The Obama Administration has also proposed "investing for performance" reforms that will improve transparency and accountability in the transportation financing system. Specific reform measures include building state and metropolitan project evaluation capabilities, improving project assessment tools, and creating stronger public reporting requirements. These reforms will promote accountability in the transportation financing system and increase the return to transportation investments, thereby boosting long-term economic growth.

- **Modernize the Electric Grid.** Our current electricity transmission grid must be expanded and modernized to reduce congestion, maintain reliability, and accommodate the output from new sources of renewable energy. New technologies are being developed that present significant opportunities for consumers and businesses to control their energy use and costs, reducing the strain on the electric grid and improving performance. The Recovery Act provides $4.5 billion for the development of technologies to enable greater energy efficiency, customer demand response, energy storage, and other components of the "Smart Grid." The Recovery Act increased the borrowing authority of the Bonneville Power Authority by $3.25 billion and provided new borrowing authority of $3.25 billion for the Western Area Power Authority, enabling both Authorities to invest in transmission lines that will increase the development of renewables in their regions.
- **Fulfill a new transportation vision with high-speed rail.** The President has proposed a long-term strategy to build an efficient, high-speed rail network of 100-600 mile intercity corridors, as one element of a modernized transportation system. The President made a down payment on this strategy with an $8 billion investment in the Recovery Act. This will be used to leverage other public and private funding to invest in infrastructure, equipment, and intermodal connections along three tracks.
- First, the strategy will advance new express high-speed corridor services (operating at speeds above 150 mph in some areas) in select corridors of 200-600 miles. Second, the strategy will develop emerging and regional high-speed corridor services (operating at speeds between 90 and 150 mph) in corridors of 100 to 500 miles. And third, the strategy will upgrade the reliability and service on conventional intercity rail services (operating at speeds up to 79-90 mph).
- The President has also budgeted for another $1 billion per year to continue to develop the high-speed rail network.
- **Develop the next generation of air traffic control.** The FY2010 Budget provides $865 million for the Next Generation Air Transportation System in the Federal Aviation Administration. The Administration supports moving from a ground-based radar

surveillance system to a more accurate satellite-based surveillance system, the development of more efficient routes through the airspace, and improvements in aviation weather information.

D. Develop an Advanced Information Technology Ecosystem

For America to lead the world in the technologies of the future, President Obama believes that all Americans must have affordable 21st century access to the Internet.

- **Expand access to broadband.** The Recovery Act provides $7.2 billion for broadband expansion and the 2010 budget includes $1.3 billion in USDA loans and grants to increase broadband capacity and telecommunication service. This support is a kickstart towards ensuring that all Americans have affordable 21st century access to the Internet. Widespread high-speed Internet access is essential for economic growth, job creation, and global competitiveness, and will foster the next generation of innovators while enabling reductions in energy consumption through telework, making online distance education tools accessible to all, enhancing remote medical monitoring capabilities, facilitating civic engagement, and supporting enhanced communications networks for first responders.
- Increased broadband access is a key input for rural economic development. It will enable rural businesses to improve their efficiency and expand their market reach; enable rural populations to compete remotely for a wide range of service jobs; and allow rural communities to retain their populations while attracting new businesses. The farm sector, a pioneer in rural Internet use, is increasingly comprised of farm businesses that purchase inputs and make sales online. Only 46 percent of adults in rural households have broadband access, compared to 67 percent for non-rural adults. America should lead the world in broadband adoption and Internet access, and we should not leave rural populations behind.
- **Assure net neutrality to preserve the freedom and openness of Internet access.** The Internet is the ultimate level playing field, making possible the widest and most lucrative variety of entrepreneurial activity and innovation that the world has ever seen. FCC Chairman Julius Genachowski announced on September 21st that providers of Internet access should not discriminate against lawful applications or content. We want to make sure it remains possible for anyone to start a business in a garage that creates new jobs, new ideas,

and new opportunities for Americans. Our nation's economy is increasingly dependent on the Internet; the Internet is an essential infrastructure, like roads and electricity; and the global leadership that America provides today stems directly from historic policies that have ensured that telecommunications networks are open to all lawful uses by all users. That's the way the Internet has always worked, and we want it to stay that way – not because we treasure our past, but because we care about our economic future.

- **Support research for next-generation information and communications technology.** The Administration is committed to supporting research that will foster the next wave of innovation in information and communications technologies, such as "cognitive radio" that allows for the efficient sharing of spectrum, quantum computing, efficient programming of parallel computers, cyber-physical systems, secure computers and networks, data-intensive supercomputers, and nanoelectronics that enables the continuation of Moore's Law for decades to come. The President's Budget supports this research through grants for the National Science Foundation, DARPA, and other public and private institutions.
- **Appoint a Chief Technology Officer of the U.S. Government**. In April, President Obama named the first ever Chief Technology Officer of the U.S. government. The CTO position was created to oversee the application of technology to create jobs and spur economic growth. The Administration is committed to recruiting high-level champions for innovation across the government.
- **Secure cyberspace**. The President has identified cybersecurity as a national priority to ensure that cyberspace – the globally-interconnected digital communications infrastructure - is sufficiently resilient and trustworthy to support U.S. goals of economic growth, civil liberties, privacy protections, national security, and the continued advancement of democratic institutions. Leadership in this effort has been anchored in the White House through the positions of the U.S. Chief Technology Officer, the U.S. Chief Information Officer, and the President's coming Cybersecurity Coordinator. This leadership team is partnering across the public and private sectors to build capacity for a digital nation through education, training and awareness; enhance security through information sharing, effective incident response planning, and privacy-enhancing identity management strategies; and encourage innovation through game-changing research and development strategies.

2. Promote Competitive Markets that Spur Productive Entrepreneurship

A. Promote American Exports

Exports will play an increasingly critical role in the future of the American economy, and the President's plans will ensure fair and open markets for American producers.

- **Open markets abroad.** The Obama Administration is working with our trading partners to negotiate mutually beneficial trade agreements and to maintain the worldwide flow of goods, services, and capital. Only by keeping markets open can American producers sell to world markets and derive the benefits of participating in an open exchange of ideas and innovations.
- **Promote American Exports.** President Obama is committed to robust support for American exporters. The United States Trade Representative (USTR) and the Department of Commerce are proactively coordinating our support through programs such as the International Trade Administration and the Trade Promotion Coordinating Committee.
- **Enforce our trade agreements to ensure access for American products abroad.** Over the last eight years the enforcement of trade agreements slowed dramatically, with the United States bringing only an average of three WTO cases per year – as opposed to the approximately 11 annually from 1995 to 2001. In this era, the United States lost its focus on ensuring that other countries lived up to their promises to open their markets, not violate America's intellectual property, and not use dumping or subsidies to penetrate America's markets. Under President Obama, USTR and the Department of Commerce are committed to a new emphasis on enforcing our existing agreements.
- **Protect intellectual property rights.** Intellectual property is to the digital age what physical goods were to the industrial age. We must ensure that intellectual property is protected in foreign markets and promote greater cooperation on international standards that allow our technologies to compete everywhere. The Administration is committed to ensuring that the United States Patent and Trademark Office has the resources, authority, and flexibility to administer the patent system effectively and issue high-quality patents on innovative intellectual property, while rejecting claims that do not merit patent protection.

- **Reform U.S. export controls**. The President has directed that the National Economic Council and the National Security Council review the overall U.S. export control system, tasking them to consider reforms that enhance America's national security, foreign policy, and economic security interests. While the U.S. has one of the most robust export control systems in the world, it remains rooted in the Cold War era of over 50 years ago. It must be updated to address the threats we face today and the changing economic and technological landscape.

B. Support Open Capital Markets that Allocate Resources to the Most Promising Ideas

Open capital markets are one of the greatest strengths of the American economy, and the President is committed to making sure these markets work.

- **Promote open capital markets.** In our economic system, freely moving capital searchers for the most promising innovations to nurture and propel. Providing American leadership in well regulated, open markets will support global technology development, and the President is committed to this system.
- **Ensure working financial markets.** While free and open markets provide significant benefits, they must work for consumers and investors. The recent financial crisis highlighted the danger of managing a 21st century economy with a 20th century regulatory framework. We are committed to building a system where individuals and businesses can innovate and take chances without fearing that the system will pose untenable risks. Our plan:
- Requires that all financial firms that pose a significant risk to the financial system at large are subjected to consolidated supervision and regulation.
- Increases supervision of financial markets to help ensure that our markets are strong enough to withstand system-wide stress and the potential failure of one or more large financial institutions.
- Rebuilds trust in our markets by creating a Consumer Financial Protection Agency to focus exclusively on protecting consumers in credit, savings, and payment markets.
- Provides the government with the tools to cope with crises by ensuring the orderly unwinding of failing firms and avoiding the untenable choice between bailouts or damaging collapse.
- Raises international regulatory standards and improves international coordination.

C. Encourage High-Growth and Innovation-Based Entrepreneurship

Entrepreneurship has played, and will continue to play, an essential role in generating innovation and stimulating U.S. economic growth. Firms with fewer than 20 employees accounted for approximately 18 percent of private sector jobs in 2006, but nearly 25 percent of net employment growth from 1992 to 2005. Small businesses employ 30% of high tech workers such as scientists, engineers, and information technology workers. The Obama Administration is committed to helping entrepreneurs build new and vibrant businesses that lead to new jobs and economic growth.

- **Increase access to capital for new businesses**. Providing access to credit for entrepreneurs and small business owners is a foundational element of economic recovery and growth. The Recovery Act reduced fees and increased guarantee levels on small business loans, while the U.S. Small Business Administration (SBA) has increased small business lending by more than 61% from the depths of the recession while broadening the base of commercial SBA lenders. Growth capital is essential. The Small Business Investment Company (SBIC) Debentures program provides debt and mezzanine financing at a time when the equity markets have pulled back from providing capital to these companies. The President has also proposed eliminating the capital gains tax on small businesses.
- **Provide training and mentoring to entrepreneurs.** Entrepreneurs with access to a network of trainers, mentors, and counselors can improve their chances for success in building high-growth businesses. The SBA has 68 district offices and over a thousand nonprofit "resource partners" that offer 14,000 counselors who serve about 1.5 million entrepreneurs and small business owners each year. The Administration is partnering with community colleges, universities, and the philanthropic sector to deliver more training and mentoring resources to aspiring entrepreneurs to promote the creation of new businesses, particularly among women and minorities.
- **Create competitive communities by promoting regional innovation clusters.** In various regions of the U.S., entrepreneurs are collaborating with local researchers, educators and industry leaders to foster specialized knowledge, technical expertise, and cutting-edge products. This will help American businesses retain and achieve new levels of competitiveness. The President's Budget provides $50 million in regional planning and matching grants within the Economic Development Administration (EDA) to support the creation of

regional innovation clusters that leverage regions' existing competitive strengths to boost job creation and economic growth. The Budget also launches a $50 million initiative in EDA that will create a national network of business incubators to encourage entrepreneurial activity in economically distressed areas.

- **Stimulate entrepreneurship through increased access to government data.** The Administration launched Data.gov, a one-stop shop for free access to data generated across all Federal agencies. By empowering the American people to find, use, and repackage data, Data.gov will give rise to new businesses (like the GPS and genomics industries that grew from increased access to public information) and empower entrepreneurs to evaluate opportunities.
- **Protect small businesses from unfair business practices**. In many industries, small companies are critical innovators, bringing enormous benefits to consumers while putting competitive pressure on incumbent firms. The Obama Administration is committed to enforcing the antitrust laws to insure that innovative entrepreneurs are not excluded from the market by anti-competitive conduct. The Department of Justice actively investigates allegations of exclusionary conduct as part of its law enforcement mission to keep markets open and competitive.

D. Improve Public Sector Innovation and Support Community Innovation

Innovation must occur within all levels of society, including the government and civil society. The Obama Administration is committed to increasing the ability of government to promote and harness innovation. The Administration is encouraging departments and agencies to experiment with new technologies that have the potential to increase efficiency and reduce expenditures, such as cloud computing. The Federal government should take advantage of the expertise and insight of people both inside and outside the Federal government, use high-risk, high-reward policy tools such as prizes and challenges to solve tough problems, support the broad adoption of community solutions that work, and form high-impact collaborations with researchers, the private sector, and civil society.

- **Make the government more transparent, participatory, and collaborative.** On his first day in office, the President signed the Memorandum on Transparency and Open Government, thereby placing government accountability and civic engagement at the forefront of the Administration's governing philosophy. The President's Memorandum urged agencies to promote three principles for bringing innovation to government: transparency, participation, and collaboration. Transparency promotes accountability by providing citizens with information about what their Government is doing. Public participation in decision-making strengthens democracy and ensures that Government makes policies with the benefit of information that is widely dispersed in society. Collaboration improves the effectiveness of Government by encouraging cooperation and knowledge-sharing within the Federal Government, across levels of Government and between the Government and private institutions.

- **Promote Open Government**. The Administration created the White House Open Government Initiative to coordinate Open Government policy, projects, and design technology platforms that foster openness across the Executive branch. The Initiative has achieved many important milestones, including:
- Publishing government data online to make it easy for anyone to remix and reuse, thus involving the American people in the development of public policy,
- Challenging thousands of Federal employees to propose ideas for slashing the time required to process veterans' disability benefits,
- Releasing information on Executive branch personnel and salaries, and
- Launching the IT dashboard, a one-stop clearinghouse of information that allows anyone with a web browser to track government spending on technology and hold the government accountable.
- **Use innovation to improve government programs.** President Obama is committed to using novel techniques and research support to improve the efficiency and effectiveness of government programs. For example, the Recovery Act includes a $7 billion fund to incentivize innovative reforms in states' Unemployment Insurance programs. States that use the most recent wage data and commit to cover more groups of job seekers get rewarded with higher payments. Already 32 states have qualified, and of these 24 of them changed their laws to do so. Another example is support for patient-centered health research in the Recovery Act. This research will lead to higher quality and more effective ways to deliver healthcare. The results will stimulate action across the health system to incorporate these findings into programs.
- **Commit White House Resources to scaling and promoting community innovations**. The President created the White House

Office of Social Innovation and Civic Participation to grow the marketplace for community innovations and provide the technology and tools for greater civic participation to help tackle our nation's toughest problems. The office will build upon efforts across the agencies, such as the Department of Education's $650 million Invest in Innovation (i3) Fund, to create new models of Federal grant-making that focus on encouraging, testing and scaling the most promising ideas and programs. The office uses its convening power to coordinate and partner with citizens, philanthropists, and the private sector to create a supportive environment for ongoing innovations in communities. Part of the effort will include using innovative tools such as prizes and challenges. The President's Budget includes $50 million in seed capital for the nation's first Social Innovation Fund, which will identify the most promising, results-oriented non-profit programs and provide the capital needed to replicate their success in communities around the country.

3. CATALYZE BREAKTHROUGHS FOR NATIONAL PRIORITIES

A. Unleash a Clean Energy Revolution

President Obama is committed to U.S. leadership in the new clean energy economy of the future. The Administration's investments will put American innovators ahead of the curve, creating new jobs in cutting edge industries while tackling the threat posed by climate change.

- **Double the nation's supply of renewable energy in the next three years.** The President has set a goal to double the generation of renewable energy in the next three years, unleashing a wave of innovation in the clean energy industries of the future. To accomplish this vision, the Recovery Act included billions of dollars to support loan guarantees and the extension of the Production Tax Credit for electricity production from renewable energy sources, leveraging tens of billions of dollars in private investment. The Recovery Act and the President's 2010 Budget also include significant increases for renewable energy technology R&D. In addition, the President pursued regulatory reform to ease the transition to a clean energy economy. In April the Department of the Interior released final regulations that will govern the development of renewable energy in offshore waters. These rules will enable our Nation to tap into the ocean's vast sustainable resources to generate clean, green energy in an environmentally sound and safe manner.

- **Promote energy efficient industries.** The Recovery Act and national energy policy will generate a proliferation of new technologies, processes, and jobs relating to improving energy efficiency. The Recovery Act alone provided $5 billion in funding for weatherization assistance to low-income residents, $4.5 billion for greening Federal buildings, and $6.3 billion for state and local government renewable energy and energy efficiency and conservation efforts.
- **Invest in Clean energy innovation**. The President has proposed a 10 year, $150 billion investment in the research, development and demonstration of clean energy technologies, such as solar, wind, green buildings, efficient lighting, next-generation biofuels, proliferation-resistant nuclear reactors, energy storage, and carbon capture and storage.
- **Enact a cap-and-trade program to curb oil dependence and greenhouse gas emissions while spurring renewable energy technologies.** President Obama supports a comprehensive cap-and-trade program that will provide a clear signal that the current energy mix is unacceptable and that low-carbon energy sources are the way of the future. The program will provide the certainty necessary for businesses to make transformative investments in renewable alternatives and energy efficiency.
- **RE-ENERGYSE the American workforce.** The President has proposed REENERGYSE, a joint educational campaign from the Department of Energy and the National Science Foundation to inspire tens of thousands of young Americans to pursue careers in clean energy. RE-ENERGYSE will support fellowships, interdisciplinary graduate programs, and partnerships between academic institutions and innovative companies to prepare a generation of Americans to meet the energy challenge.

B. Support Advanced Vehicle Technologies

For the sake of our national security, economy, and environment, it is crucial to diversify away from oil as the sole source of transportation fuel. Today, oil accounts for 96 percent of the fuel that powers our country's vehicle fleet. The lack of widespread, affordable alternatives to oil makes us vulnerable to market disruptions and dependent on oil producers. Oil is also responsible for one third of our carbon dioxide emissions and is a significant source of local air pollution, threatening our climate security and the health of our local communities.

The President's strategy is to put the U.S. at the cutting edge of the advanced vehicle technology industry, which will not only reduce our dependence on oil, but will also create jobs, strengthen our manufacturing base, improve the quality of the air we breathe, and offer consumers greater safety, performance, and choice.

- **Make the largest investment in technology for electric vehicles and transportation electrification in U.S. history.** In early August the Administration announced $2 billion in grants which will catalyze private sector investment to build a globally competitive domestic battery and electric drive component industry. With this support, American factories will produce the lightest, cheapest, longest-lasting, and most powerful vehicle batteries in the world. American batteries and components will power affordable electric cars that can travel over hundred miles on a single charge and offer customers superb performance. These vehicles will be even more attractive thanks to the tax credit of up to $7,500 offered for electric and plug-in electric vehicles in the Recovery Act.
- The Administration is also making a $400 million holistic investment in transportation electrification as a system. Beyond batteries, vehicles, and components, this includes demonstrating pilot systems that put the pieces together – testing the infrastructure to plug the cars, training the workers to build and service them, and educating the consumers who will buy them. The lessons from these projects will help the private sector more quickly develop profitable electric vehicles that meet the needs of drivers.
- **Deploy up to $25 billion in loans to support American manufacturing of advanced vehicle technologies.** Through the $25 billion Advanced Technology Vehicles Manufacturing Loan Program, the Administration is supporting competition within the marketplace to produce the most cost-effective solutions to reduce oil dependence. The Administration awarded the first $8 billion in conditional loan commitments in June.
- These included $5.9 billion for Ford Motor Company to transform factories across Illinois, Kentucky, Michigan, Missouri, and Ohio to produce 13 more fuel efficient models; $1.6 billion to Nissan North America, Inc. to retool their Smyrna, Tennessee factory to build advanced electric automobiles and to build an advanced battery manufacturing facility; and $465 million to Tesla Motors to manufacture electric drive trains and electric vehicles in California. Up to an additional $17 billion in loans will be made under this ... over the next several years

- **Support the next generation of American biofuels.** The Administration is investing in next generation biofuels that displace oil consumption and reduce greenhouse gas emissions. Through $800 million in Recovery Act grants as well as up to $500 million to support loan guarantees, the Administration is accelerating the development of clean technologies like cellulosic and algae-based biofuels –harnessing recent advances in synthetic biology.
- **Improve vehicle fuel efficiency to reduce oil dependence and spark innovation.** In May, President Obama announced a groundbreaking national autos program that put us on a path to adopting uniform Federal standards to regulate both fuel economy and greenhouse gas emissions. The result will be a projected reduction in oil consumption of approximately 1.8 billion barrels over the life of the program and a projected total reduction in greenhouse gas emissions of approximately 950 million metric tons. This landmark policy will spark innovation in more fuel-efficient vehicles, reduce pollution, and promote energy security.

C. Drive Innovations in Health Care Technology

The inefficiencies in our health care system raise costs and reduce the quality of care. New advances in health information technology will increase efficiency while broad reform will free businesses and individuals to innovate and grow.

- **Expand the use of health IT.** Expanded use of advanced health information technology (e.g. electronic medical records, mobile health applications, sensors for monitoring chronic diseases) will help prevent medical errors, improve health care quality, begin to modernize the American health care system and reduce costs. The Recovery Act provides over $19 billion in investments to modernize health information technology.
- **Renew our commitment to medical research.** The Recovery act included a $10 billion expansion in health research. This will fund projects such as an initiative to identify all of the genetic changes involved in 20 types of cancer, clinical trials of medicines that could help stop the HIV/AIDS pandemic, the largest infusion of funding to discover the causes and treatment for autism, and using DNA sequencing to discover how to prevent and treat heart, lung, and blood diseases

- **Slow the growth of health care costs.** The President is committed to comprehensive reform for a health care system that makes it possible to improve the quality of care while slowing the growth rate of costs. Doing this will free up resources that can be used to invest in American businesses and improve the living standards for all Americans.

D. Harness Science and Technology to Address the "Grand Challenges" of the 21st Century

The President's renewed commitment to science, technology and innovation will allow the nation to set and meet ambitious goals that will improve our quality of life and establish the foundation for the industries and jobs of the future. Examples include:

- Complete DNA sequencing of every case of cancer; smart anti-cancer therapeutics that kill cancer cells and leave their normal neighbors untouched; early detection of dozens of diseases from a saliva sample; nanotechnology that delivers drugs precisely to the desired tissue; personalized medicine that enables the prescription of the right dose of the right drug for the right person; a universal vaccine for influenza that will protect against all future strains; and regenerative medicine that can end the agonizing wait for an organ transplant.
- Solar cells as cheap as paint, and green buildings that produce all of the energy they consume.
- A light-weight vest for soldiers and police officers that can stop an armor-piercing bullet.
- Educational software that is as compelling as the best video game and as effective as a personal tutor; online courses that improve the more students use them; and a rich, interactive digital library at the fingertips of every child.
- Intelligent prosthetics that will allow a veteran who has lost both of his arms to play the piano again.
- Biological systems that can turn sunlight into carbon-neutral fuel, reduce the costs of producing anti-malarial drugs by a factor of 10, and quickly and inexpensively dispose of radioactive wastes and toxic chemicals.
- An "exascale" supercomputer capable of a million trillion calculations per second – dramatically increasing our ability to understand the world around us through simulation and slashing the time needed to

design complex products such as therapeutics, advanced materials, and highly-efficient autos and aircraft.

- Automatic, highly accurate and real-time translation between the major languages of the world – greatly lowering the barriers to international commerce and collaboration.

Strategies for American Innovation
Editor: Zachary A. Monroy
ISBN: 978-1-61324-247-6

Chapter 3

STATEMENT OF ANEESH CHOPRA, CHIEF TECHNOLOGY OFFICER AND ASSOCIATE DIRECTOR, OFFICE OF SCIENCE AND TECHNOLOGY POLICY, EXECUTIVE OFFICE OF THE PRESIDENT OF THE UNITED STATES, TO THE COMMITTEE ON COMMERCE, SCIENCE AND TRANSPORTATION, SUBCOMMITTEE ON COMPETITIVENESS, INNOVATION AND EXPORT PROMOTION, UNITED STATES SENATE ON INNOVATION IN AMERICA: OPPORTUNITIES AND OBSTACLES JUNE 22, 2010

Chairman Klobuchar, Ranking Member LeMieux, and Members of the Subcommittee, it is my distinct privilege to be here with you today to discuss the Obama Administration's Strategy for American Innovation.

President Obama understands the importance of innovation for sustainable growth and quality jobs. On September 21st, 2009, he released his *Strategy for American Innovation* that identified three critical roles for the Federal Government: to invest in the building blocks of innovation; to create the right environment for private sector investment and competitive markets by, for example, promoting high-growth entrepreneurship, protecting U.S. intellectual property rights, and fostering an open government; and to serve as a catalyst for breakthroughs related to national priorities such as clean energy, health care, and other "grand challenges" of the 21st century.

In my capacity as Assistant to the President, Chief Technology Officer, and Associate Director for Technology in the Office of Science and Technology Policy, my mission is to harness the power and potential of technology, data, and innovation to transform the Nation's economy and to improve the lives of everyday Americans. The Administration envisions an economy in which jobs are more plentiful, American firms are more competitive, Americans are safer and more secure, and energy use is cleaner and more economical.

PROBLEMS WITH THE BUBBLE-DRIVEN GROWTH OF THE PAST

Despite the American economy's historic strength, our economic growth has rested for too long on an unstable foundation. Time and again, explosive growth in one sector of our economy provided a short-term boost while masking long-term weaknesses. In the 1990s, the technology sector climbed to unprecedented heights of valuation. The tech-heavy NASDAQ composite index rose over 650 percent between 1995 and 2000, but then lost two-thirds of its value in a single year.

After the tech bubble burst, a new one emerged in the housing and financial sectors. This type of growth isn't just problematic when the bubble bursts, it is not entirely healthy even while it lasts. Between 2000 and 2007 the typical working-age American household saw its annual income decline by nearly $2,000.

A short-term approach to the economy masks under-investments in essential drivers of sustainable, broadly-shared growth. It promotes temporary fixes over lasting solutions. This is patently clear when looking at how American education, infrastructure, healthcare, energy, and research – all pillars of lasting prosperity – were ignored during the last bubble.

Despite this underinvestment in key drivers of growth, the American economy remains the most dynamic, innovative, and resilient in the world. America's strengths are clear: world-class research universities, flexible labor markets, deep capital markets, and an energetic entrepreneurial culture. The United States must redouble its efforts to give our world-leading innovators every chance to succeed. America cannot rest on our laurels while other countries are catching up.

The Need for Innovation

Innovation is at the core of a new foundation for durable, sustainable expansion in both employment and economic growth. Robert Solow won the Nobel Prize in economics by showing that factors other than capital intensity, most notably advances in human knowledge and technology, accounted for almost 90 percent of the growth in America's output per hour in the first half of the last century. Growth accounting has been refined since Solow's first attempts, yet contemporary research still shows that human skill and innovation remain far and away the most powerful force for improving prosperity over the long-run, which is exactly what we need.

Given its importance, the process of innovation cannot be taken for granted. Innovation begins from scientific research that creates new opportunities for technological change. That basic research lays the groundwork for the development of new products, services, or processes. But it does not end there. To create value, a new idea must be implemented. Thus successful innovations will diffuse throughout an economy and across the world, impacting various sectors and sometimes even creating new ones. A diffused innovation must then scale appropriately, reaching an efficient size at which it can have a maximal effect.

The full process -- from development to diffusion to scaling -- has many variables and many inputs. Ideas often fail before they make it through the full chain. But those that do succeed can create value and jobs while improving people's lives.

For societies to prosper—both as producers of goods and services as well as consumers of them—innovations need to flourish and progress along this chain. And here, government has a fundamental role to play.

The Appropriate Role for Government

While it is clear that a new foundation for innovation and growth is needed, the appropriate framework for government involvement is still debated. For the Obama Administration, the arguments about too much or too little government involvement in innovation policy often lead to unproductive debates. The real question is how can government best create the conditions that will enable private sector entrepreneurs to innovate. Stated differently, the real issue is how to enable entrepreneurs to move our economy forward. As explained in the Innovation Strategy document, the best way forward is for the United States to invest in the building blocks that only the government can provide, protecting an open and competitive environment for businesses and individuals to experiment and grow, and by providing extra catalysts to jumpstart innovation in sectors of national importance.

A Strategy for American Innovation

President Obama has already taken historic steps to lay the foundation for the innovation economy of the future. In the Recovery Act alone, the President committed over $100 billion to support groundbreaking innovation with investments in energy, basic research, education and training, advanced vehicle technology, health IT and health research, high speed rail, smart grid, and information technology.

The Obama Innovation Strategy has three parts: investing in the building blocks of innovation, promoting competitive markets that spur productive entrepreneurship, and catalyzing breakthroughs for national priorities.

Investing in the Building Blocks of American Innovation

President Obama is committed to making investments that will foster long-term economic growth and productivity. These investments are in areas that include research and development, a skilled workforce, a leading physical infrastructure, and widely available broadband networks.

Recognizing the need for long-term and sustained investments in R&D, President Obama has pledged to complete the doubling of funding for three key science agencies, the National Science Foundation, the laboratories of the National Institute of Standards and Technology, and the Department of Energy's (DOE) Office of Science. In his landmark address before the National Academy of Sciences, President Obama set a goal of lifting the sum of public and private investment in R&D to 3 percent of GDP, which would exceed the level achieved at the height of the space race. As the President noted, "science is more essential for our prosperity, our security, our health, our environment and our quality of life than it has ever been before."

To encourage private sector investment in R&D, the President has proposed making the Research and Experimentation Tax Credit permanent. The Obama Administration is working to increase the impact of this investment by providing greater support for university commercialization efforts, for high-risk, high-return research, for multidisciplinary research, and for scientists and engineers at the beginning of their careers. For example, the National Science Foundation's FY11 budget proposes to double support for the Partnerships for Innovation program, which will help universities move ideas from the lab to the marketplace.

The Obama Administration is committed to expanding access to broadband. This past April, the Federal Communications Commission (FCC) released the National Broadband Plan, called for in the American Recovery and Reinvestment Act, to identify ways to expand access to broadband and promote economic growth and job creation.

In his statement on the plan's release, the President committed to "build upon our efforts over the past year to make America's nationwide broadband infrastructure the world's most powerful platform for economic growth and prosperity." To that end, I've established a Broadband Subcommittee of the National Science and Technology Council's Committee on Technology, to focus closely on the plan that the FCC—an independent agency—produced, and to advise the Administration on the actions it should take to promote broadband as a platform to improve the lives of everyday Americans and drive innovation in the economy.

Promoting Competitive Markets that Spur Productive Entrepreneurship

The Obama Administration believes that it is imperative to create a national environment that is ripe for entrepreneurship and risk taking, and allows U.S. firms to compete and win in the global marketplace. The Administration is pursuing policies that will promote U.S. exports, support open capital markets, encourage high-growth entrepreneurship, invest in regional innovation clusters, and improve our patent system. The Administration also strongly supports public sector and social innovation.

Competitive, high-performing regional economies are the building blocks for national growth and job creation, and the Administration is stepping up its efforts to cultivate regional economic clusters across the country. For example, the Administration recently announced that seven federal agencies would work together on a $130 million competition for an Energy Regional Innovation Cluster (E-RIC) around one of DOE's Energy Innovation Hubs. The

Department of Commerce's Economic Development Administration is one of the partners, and will be contributing funds to link the Hub with local economic development strategies and to support economic adjustment efforts in the local community. This pilot project is designed to spur regional economic growth while developing energy efficient building technologies, designs, and systems. This will allow a region to develop a strategy that includes support for R&D, infrastructure, small and medium-sized enterprises, and workforce development. What we are learning is that whether the investment comes from the federal or state government, or the private sector, or-ideally-all of the above, those dollars will do a lot more good if they serve a well-developed regional strategy that leverages core regional strengths.

The i6 Challenge launched by the Commerce Department's Economic Development Administration, the National Institutes of Health, and the National Science Foundation, is another example of these efforts. A total of $12 million is available to six teams around the country with the most innovative ideas to drive technology commercialization and entrepreneurship in their regions. The i6 Challenge is aimed at bringing together entrepreneurs, investors, universities, foundations and non-profits in communities throughout the United States, with applications due by July 15, 2010.

Innovation must occur within all levels of society, including the government and civil society. The Obama Administration is committed to increasing the ability of government to promote and harness innovation. The Administration is encouraging departments and agencies to experiment with new technologies that have the potential to increase efficiency and reduce expenditures, such as cloud computing. The Federal Government should take advantage of the expertise and insight of people both inside and outside the Federal government; use high-risk,

high-reward policy tools such as prizes and challenges to solve tough problems; support the broad adoption of community solutions that work; and form high-impact collaborations with researchers, the private sector, and civil society.

The Administration launched the White House Open Government Initiative to coordinate Open Government policy, support specific projects, and design technology platforms that foster transparency, participation and collaboration across the Executive Branch. The principles of open government help to advance a set of key national priorities with emphasis on demonstrating tangible benefits for the American people.

An example of how prizes are being used to spur national priorities is USDA's Apps for Healthy Kids challenge that was launched by First Lady

Michelle Obama in March as part of her Let's Move initiative. The competition is based upon a recently released set of data on nutrition by USDA and is aimed at encouraging entrepreneurs, software developers and students to create applications and games that encourage children and parents to make more nutritious food choices and to be more physically active. Eight game jams, bringing together developers to share tips and ideas, have been held across the country, and over twenty applications have been submitted so far in advance of the contest deadline on June 30th.

Catalyzing Breakthroughs for National Priorities

President Obama is committed to harnessing science, technology and innovation to unleash a clean energy revolution, improve America's health care system, and address the "grand challenges" of the 21st century.

Smart Grid Technologies

Modernization of the Nation's electric grid is a vital component of efforts to build a low-carbon economy. The "smart grid" will help provide consumers with the information, automation, and tools they need to control and optimize energy use. The tools and services enabled by the smart grid promise improve the reliability, security, and efficiency of the electric grid. Smart grid technologies can also facilitate energy generation from clean energy supplies and enable more effective integration with the electricity delivery system of renewable energy sources, demand response resources, and plug-in electric vehicles. The National Institute of Standards and Technology (NIST) has coordinated an unprecedented, open and transparent public/private collaboration involving over 550 companies, organizations and government agencies to create the interoperability standards needed to foster innovation in the electric grid.

Last March, in conjunction with NIST, we broadened participation by launching the Smart Grid Forum, an on-line forum focused on the Nation's energy consumers with an emphasis on spurring innovation in smart grid products and services. We received comments from over 130 individuals and organizations contributing their solutions to some of the most challenging smart grid goals that we have – from deployment of smart grid solutions, to development of standards needed for information exchange, to ensuring cybersecurity in the smart grid. Following the input received in that forum, I established another Subcommittee of the National Science and Technology

Council's Committee on Technology to enable the Administration to develop a comprehensive policy framework for Smart Grid policy.

Healthcare IT

Another important Presidential priority is improving our health care system. Broad use of health information technology has the potential to improve health care quality, prevent medical errors, increase the efficiency of care provision and reduce unnecessary health care costs, reduce paperwork, increase administrative efficiencies, expand access to affordable care, and improve population health. The Recovery Act provides support for the deployment of health information technology, such as electronic health records. The Office of the National Coordinator for Health IT and the Centers for Medicare & Medicaid Services are working to ensure that health information technology products and systems are secure, can maintain data confidentially, can work with other systems to share information, and can perform a set of well-defined functions. NIST, in coordination with the Office of the National Coordinator and others, is accelerating the adoption of health IT standards by providing the critical testing infrastructure needed to achieve these goals.

Last February, the Office of the National Coordinator for Health IT announced a new collaborative, *NHIN Direct*, which will organize a set of standards, services and policies that enable secure health information exchange over the Internet (www.nhindirect.org). Several Federal agencies and healthcare organizations are already using the Nationwide Health Information Network (NHIN) technology to exchange information amongst themselves and their partners. This new effort will provide an easy "on-ramp" for a wide set of providers and organizations looking to adopt the exchange of health information – and provide a framework to spur innovation in support of direct communication amongst providers, and between providers and patients – in a secure and simple manner.

The recently launched Community Health Data Initiative (CHDI) is another effort demonstrating how data and the innovative uses of technology are resulting in immediate improvements to health care. A public-private collaboration spearheaded by the Department of Health and Human Services, CHDI is aimed at using health care data to raise awareness and improve community health performance. Innovators from the worlds of business, technology, academia, and community organizations identified areas where exciting new applications to improve health could be developed. In less than 12 weeks these partners put together an amazing array of new or improved applications that utilize our data in creative and powerful ways to help advance health care. The results of these efforts, unveiled in early June, included the integration of patient satisfaction ratings from Medicare's Hospital Compare database into the web search results for hospitals, and a brilliant new combination of GPS device and app that allows asthmatics to have their inhalers automatically transmit the location and time of each use – producing an anonymized, real-time map of asthma incidence that can provide crucial guidance regarding how to target interventions to reduce the burden of asthma.

Existing technologies are also being used in innovative ways to improve health education, through the Text4Baby campaign that was launched in February. Text4Baby is a free mobile health education service to promote maternal and child health. Expecting mothers can text baby, or bebe in Spanish, to 511411, and they will receive free SMS text messages each week, timed to their due date or their baby's date of birth. Fifteen wireless carrier have agreed to deliver Text4Baby messages to subscribers at no charge for two years, and as a result, nearly 50,000 individuals have signed up for this services since February.

Grand Challenges

Finally, the Obama Administration believes that grand challenges should be an important organizing principle for science, technology and innovation policy. They can address key national priorities, catalyze innovations that foster economic growth and quality jobs, spur the formation of multidisciplinary teams of researcher and multi-sector collaborators, bring new expertise to bear on important problems, strengthen the "social contract" between science and society, and inspire students to pursue careers in science, technology, engineering, and mathematics. The President's innovation strategy sets forth a number of grand challenges, such as solar cells as cheap as paint, educational software that is as compelling as the best video game and effective as a personal tutor, and early detection of diseases from a saliva sample. The National Economic Council and the Office of Science and Technology Policy are encouraging multi-sector collaborations to achieve these grand challenges that might involve companies, research universities, foundations, social enterprises, non-profits, and other stakeholders.

The Way Forward

Thanks to President Obama's leadership, the Administration has taken large strides in developing and implementing an ambitious innovation agenda. The Recovery Act alone provides over $100 billion to support research and development and the deployment of advanced technologies such as clean energy, health IT, the smart grid, and high-speed rail. This commitment to investing in America's future continues in the President's most recent budget, with sustained support for research, entrepreneurial small businesses, education reform, college completion, and a 21st century infrastructure.

The Obama Administration believes that the America COMPETES Act should be reauthorized this year so that the Nation can continue to build on the achievements of the original Act. I share the belief that the President and the Vice President hold, who supported the original COMPETES Act when they were Senators, that the COMPETES Act provides a valuable roadmap to guide Federal policies in innovation, competitiveness, and STEM education. We are supportive of this Committee's efforts to reauthorize this landmark act this year, and we very much look forward to working with the Committee to make the reauthorization a reality during this session of Congress.

The Administration is working with a wide range of stakeholders to identify the most promising ideas for implementing and further refining the Administration's innovation strategy. There are active inter-agency working groups on issues such as prizes and challenges, regional innovation clusters, research commercialization, spectrum reform, broadband, open government, and standards. The National Science and Technology Council is leading multi-agency research initiatives in dozens of critical areas such as aeronautics, genomics, green buildings, nanotechnology, quantum information science, robotics, and information technology. Through the President's Council of Advisors on Science and Technology, the Administration is able to receive high quality advice from the Nation's leading scientists, engineers and innovators on issues such as health information technology, advanced manufacturing, clean energy, and STEM education.

America has always been a Nation built on hope – hope that we can build a prosperous, healthy world for ourselves and for our children. These long-standing American aspirations depend critically on our far-sighted investments in science, technology and innovation that are the ultimate act of hope and will create the most important legacies we can leave.

The United States is still the land of the future. We have held that honor since this continent was discovered by a daring act of exploration more than

500 years ago. We have earned it anew with each passing generation because America's scientists, entrepreneurs and public officials have understood the importance of applying the power of American curiosity and ingenuity to the biggest economic and societal challenges.

I welcome any questions that the Committee may have.

INDEX

C

F

G

H

I

J

K

L

M

Q

R

S

T

U

V

W

Y